Stuck in Transition

DIRECTIONS IN DEVELOPMENT
Energy and Mining

Stuck in Transition

*Reform Experiences and Challenges Ahead
in the Kazakhstan Power Sector*

Mirlan Aldayarov, Istvan Dobozi, and Thomas Nikolakakis

WORLD BANK GROUP

Contents

Figures

Maps

Tables

Acknowledgments

This study was prepared by a World Bank team led by Mirlan Aldayarov (senior energy specialist), which included Istvan Dobozi (lead energy economist, consultant), Aksulu Kushanova (consultant), Raphael Torquebiau (consultant), and Dung Kim (program assistant) and the system modeling group: Debabrata Chattopadhyay (senior energy specialist), Thomas Nikolakakis (energy specialist), and Rhonda Jordan (energy specialist). Substantial field data gathering efforts were made by Darkhan Kurmanbayev and Dastan Sadvakassov. Important inputs were provided by World Bank staff including Vivien Foster (lead economist, global lead, and peer reviewer), Husam Mohamed Beides (lead energy specialist and peer reviewer), Sunil Khosla (lead energy specialist), Kari Nyman (lead energy specialist and peer reviewer), and Sarosh Sattar (senior economist). Substantial macroeconomic sections were drawn on work carried out by Dorsati Madani (senior economist) and Ilyas Sarsenov (senior economist). The team acknowledges its strong partnership with the team from the International Finance Corporation: Efstratios Tavoulareas (senior operations officer) and Pedro Robiou (senior energy specialist). Strong management guidance and substantial technical and policy inputs were received from Ranjit Lamech (practice manager). Management review was provided by Ludmila Butenko (former country manager for Kazakhstan) and Francis Ato Brown (country manager for Kazakhstan). The report was edited by Margie Peters-Fawcett and Amy Gautam.

During the preparation of this study, the team benefited from meetings with the Kazakhstan Ministry of Energy, Kazakhstan Electricity Grid Operating Company, Samruk-Energy, and the Kazakhstani Operator of Electric Energy and Capacity. Least-cost modeling analysis drew on earlier studies by DNV GL (formerly KEMA), funded by the European Bank for Reconstruction and Development, and the KazEnergy Association's 2013 "KazEnergy National Energy" report. An advanced draft of the study was discussed in Astana with the Kazakhstan Ministry of Energy, the Kazakhstan Electricity Grid Operating Company, Samruk-Energy, Kazakhstani Operator of Electric Energy and Capacity, and others in April–May 2016.

The study has been funded by the Energy Sector Management Assistance Program and the Central Asia Energy-Water Development Program.

The Energy Sector Management Assistance Program is a global knowledge and technical assistance trust fund administered by the World Bank that assists low- and middle-income countries to increase know-how and institutional capacity to achieve environmentally sustainable energy solutions for poverty reduction and economic growth.

The Central Asia Energy-Water Development Program is a knowledge and technical assistance multidonor trust fund administered by the World Bank, with a mission to build energy and water security for the countries of Central Asia—Kazakhstan, Kyrgyz Republic, Tajikistan, Turkmenistan, and Uzbekistan—through enhanced regional cooperation. Since its inception in 2010, the program has received support from bilateral and multilateral donors, including the Government of Switzerland's State Secretariat for Economic Affairs, the European Commission, the United Kingdom's Department for International Development, the U.S. Agency for International Development, and the World Bank Group.

Abbreviations

AREM	Agency for Regulation of Natural Monopolies
CAPEX	capital expenditure
CAPS	Central Asia Power System
CASA-1000	Central Asia-South Asia (1,000 MW)
CCGT	combined cycle gas turbine
CHP	combined heat and power plant
CO_2	carbon dioxide
EE	energy efficiency
ESO	electricity supply organization
EU	European Union
FSC	Financial Settlement Center
FT	fixed tariff
GCal	gigacalories
GDP	gross domestic product
GJ	gigajoule
GoK	government of Kazakhstan
GRES	Regional State Power Station
GW	gigawatt
GWh	gigawatt hour
HPP	hydropower plant
IEA	International Energy Agency
KazEnergy	Energy Association of Kazakhstan
KEGOC	Kazakhstan Electricity Grid Operating Company
KOREM	Kazakhstan Operator of Electric Energy and Capacity
KW	kilowatt
KWh	kilowatt hour
KZT	Kazakhstan tenge
LCOE	levelized cost of electricity
mtoe	million tons of oil equivalent

MW	megawatt
MWh	megawatt hour
m^3	cubic meters
O&M	operations and maintenance
OCGT	open cycle gas turbine
OECD	Organisation for Economic Co-operation and Development
OPEX	operating expenditure
PPA	power purchase agreement
PV	photovoltaic
RE	renewable energy
REC	regional electricity distribution company
SAIDI	System Average Interruption Duration Index
SAIFI	System Average Interruption Frequency Index
SBM	single buyer model
SEB	state electricity boards
toe	ton of oil equivalent
TPP	thermal power plant
TWh	terawatt hour
VAT	value-added tax
VRE	variable renewable energy

CHAPTER 1

Introduction

Background

The large-scale transformation of Kazakhstan's power sector after independence in 1991 was reflected by the country's move toward liberalizing the market and implementing sector regulation. Kazakhstan's power sector was an early adopter of a liberalized multimarket model—consisting of bilateral, spot, balancing, ancillary, and capacity submarkets. The sector was regarded as a market reform leader among countries of the former Soviet Union, having achieved much improved supply and demand balance and service quality. The wholesale electricity market was liberalized and operated mainly on the basis of bilateral contracts between generators and large consumers and regional electricity distribution companies for direct sale of power. The government of Kazakhstan established the legislative, technical, and organizational infrastructure for a functioning electricity spot market, which increasingly supplemented bilateral contracts as a liquid trading floor for short-term transactions. However, despite the noteworthy headway, the sector reforms remain predominantly as unfinished business. The excess generation capacity that was inherited from the former Soviet Union—at a time when "energy-only" market prices were too low to attract serious investors—has masked the need to reflect on the long-term outlook of the country's power production.

As the investment crunch unfolded in the mid-2000s, a diverging concern almost immediately arose; that is, the additional capacity of existing and planned generation may not be sufficient to keep pace with the ongoing and significant increase in the demand for power. Instead of applying market mechanisms to allow prices to rise and reflect the underlying supply and demand gap, the government of Kazakhstan addressed the issue by implementing administrative, command-and-control measures that included the introduction of energy generation tariff regulation, renationalization and oligopolization of power generation, restrictions on electricity spot market transactions, elimination of zonal transmission tariffs, and postponement of the real-time balancing market.

This study draws on the World Bank's long-standing engagement in Kazakhstan's energy sector and several recent technical assistance and advisory support activities.

Although the World Bank's engagement in Kazakhstan's energy sector has a long track record, it has not yet developed a comprehensive sector assessment report. The study aims to (a) inform the government of Kazakhstan on policy making, by objectively identifying the principal challenges faced by the power sector in its ongoing transition and outlining potential policy options; and (b) draw lessons from Kazakhstan's experience in sector reforms for the broader international audience. On the basis of extensive analysis, detailed interviews, and system modeling, the study aims to model various sector development scenarios, quantify their costs and benefits, identify key sector challenges, and recommend policy actions going forward. The study covers broad sector issues, including long-term, least-cost power system planning; supply and demand balancing; tariff setting; market structure; and integration of renewable energy. Although the focus is on the power system, the study also includes combined heat and power plants, given their considerable share in power generation.

Key Sector Challenges

Kazakhstan's power sector faces several challenges, outlined in chapter 6. These have been aggravated by the recent plunge in world commodity prices and the consequent reduction of industrial production and relative power demand. The key challenges to address are (a) supply security risk, (b) the overwhelming need for investment; and (c) the need for efficient regulation and continuous reform.

- *High energy intensity and generation capacity tightness, with the insufficient reserve cushion posing supply security risks.* The energy intensity of the gross domestic product (GDP) is very high in international comparison, placing heavy demand pressure on the power infrastructure. The reserve margin reached a high of 53 percent in 2000, when peak energy demand reached its lowest levels since the breakup of the former Soviet Union. Since then, the capacity margin has rapidly and steadily shrunk to the dangerously low level of 4 percent in 2012.
- *Formidable investment requirements.* Undiscounted annualized capital expenditure requirements, ranging from US$54.6 billion (Least-Cost Case scenario) to US$96.2 billion (Green Case scenario) over the period 2015–45, are required to meet the growing projected demand for power. The annual investment need is equivalent to 0.8–1.4 percent of the 2013 GDP. The levelized cost of energy (LCOE)[1] requires a 40–55 percent increase in the average residential tariff, although an affordability analysis confirms that electricity is generally affordable.
- *Ineffective regulation and sector reform reversal.* The government of Kazakhstan's administrative decisions since the mid-2000s have effectively rolled back several market-oriented reforms. Although at times these actions have resulted in short-term gains, the rollback has substantially aggravated longer-term prospects by worsening the investment climate, damaging competition, and crowding out private initiatives.

These challenges are interlinked. Although each challenge may call for a tailored set of solutions, some of the recommended solutions cut across challenges.

Approach and Methodology

On the basis of extensive analysis, interviews with counterparts in Kazakhstan, and system modeling, this study aims to model several plausible sector development scenarios, quantify their costs and benefits, identify key sector challenges, and offer forward-looking policy recommendations. Most of the findings and recommendations are based on the data available from 2013 to 2014.

The system modeling analysis provides an updated and refined view of Kazakhstan's capacity and generation mix for the period 2015–45, and informs decisions on the selection of various alternative generation technologies and their sizing and sequencing. A long-term, least-cost investment study used the Power System Research (PSR) planning software, operated by the Power Systems Planning team of the World Bank's Energy and Extractives Global Practice. The PSR software was developed by PSR Inc., a global provider of technological solutions and consulting services in the areas of electricity and natural gas since 1987. The analysis uses data from earlier studies with supply and demand projections and sector information to create a mathematical power system model that is updated with recent, actual numbers and current capital and operational expenditure cost estimates. Four scenarios are modeled:

- *Base Case scenario.* The most likely scenario, the Base Case scenario optimizes generation and transmission while considering existing policies, goals, and investment projects in the process of being implemented or likely to be implemented.
- *Green Case scenario.* The Green Case scenario optimizes a path toward green growth, as described in the concept for Kazakhstan's transition to a green economy, approved by the Decree of the President in May 2013 (Green Economy Concept). The scenario aims to identify the power sector's economic costs and benefits that are associated with an aggressive energy efficiency program to substantially reduce growth in demand (especially peak demand).
- *Regional Export Case scenario.* This scenario shows the economic benefits and costs if Kazakhstan were to invest in additional capacity to increase gradually its export activities while maintaining full electricity independence. Full electricity independence is a key objective of the "Concept of Development of the Fuel and Energy Complex of Kazakhstan till 2030" (referred to as Energy Concept 2030), which was approved by the government of Kazakhstan in June 2014.
- *Least-Cost Case scenario.* As an extreme benchmark, the Least-Cost Case scenario optimizes system capacity expansion and operation, purely on least-cost principles, without imposing policies or targets (which are implemented only if they are found to be economical). A sensitivity analysis estimates the impact of using the higher economic cost of natural gas (export price as a proxy) instead of the lower actual price. The main findings are shown in table 1.1.

Most of the findings and recommendations of this study are based on data available from 2013 to 2014. Hence, the latest macroeconomic developments—including the drop in oil prices and the large depreciation of the tenge—are not

Table 1.1 Comparison of the Four Modeled Scenarios

Measure	Base Case	Green Case	Regional Export Case	Least-Cost Case
System-wide levelized cost of electricity (US$/megawatt hour (MWh))	35.1	41.5 (33)[a]	34.8	31.1 (34.6)[b]
Total undiscounted annualized capital expenditure (US$ billions)	81.56	96.2	83.36	54.62 (50.8)
Total discounted annualized capital expenditure (US$ billions)	25.3	28.9	25.5	17.4 (15.2)
Total operational cost (US$ billions)	85.3	82.4	93.6	91.1 (104.9)
Average operational cost (US$/MWh)	19.2	20.4	19.7	20.6 (23.2)
Total emissions (million tons of CO_2)	2,932	2,460	3,252	2,977 (3,400)
Average emissions intensity (tons of CO_2 per megawatt hour of electricity)	0.69	0.64	0.71	0.69 (0.8)
CO_2 reductions from 2012 levels by end of planning period (%)	n.a.	40	n.a.	n.a.

Source: World Bank calculations.
Note: CO_2 = carbon dioxide; MWh = megawatt hour; n.a. = not applicable.
a. The value in parentheses is the levelized cost of energy if the benefits of global externalities associated with carbon dioxide savings were considered.
b. The values in parentheses represent the results for the variation of the Least-Cost Case scenario, which considers the economic cost of natural gas.

fully reflected. Although these developments may carry some short- to medium-term impacts, the longer-term outcomes would not be substantially influenced, so the key findings emerging from the scenario modeling are expected to remain valid.

The key challenges, findings, and recommendations of this study are presented in chapter 6.

Note

1. Traditionally, the LCOE is an economic assessment of the average total cost to build and operate a power-generating asset over its lifetime, divided by the total energy output of the asset over that lifetime. In this study, the "systemwide" LCOE is a similar concept that represents the average total cost to build, rehabilitate, and operate systemwide generation assets and interzonal high-voltage transmission over the specified planning horizon, divided by the total energy output of the system over that same horizon. Therefore, in this study, systemwide LCOE excludes transmission and distribution, but captures generation assets and the costs of the few interzonal interconnections under consideration.

CHAPTER 2

Country Context and Economic Outlook

Kazakhstan is the world's largest landlocked country and the ninth-largest country, with a land area of 2,724,900 square kilometers. It borders China, the Kyrgyz Republic, the Russian Federation, Turkmenistan, and Uzbekistan, and adjoins a large part of the Caspian Sea. With 17.5 million people as of 2014, Kazakhstan is the 61st-most-populous country in the world, although its population density is among the lowest, at fewer than six people per square kilometer.

Kazakhstan has the second-largest oil reserves among the former Soviet Union countries, as well as the second-largest oil production, after Russia (figures 2.1 and 2.2). Total oil production was 1.7 million barrels a day in 2014, and further growth is contingent on the timely development of the giant Tengiz, Karachaganak, and Kashagan fields.

Rising natural gas production over the past decade has boosted oil recovery (as a significant volume of natural gas is reinjected into oil reservoirs) and decreased Kazakhstan's reliance on natural gas imports. However, gas consumption has been stagnant, because the infrastructure and expense required to connect Kazakhstan's widely dispersed population to production centers in the northwest have impeded gas penetration. Kazakhstan is landlocked and located far from major international oil markets. The country depends mainly on pipelines to transport its hydrocarbons to export markets. Kazakhstan is a transit country for natural gas exports from Turkmenistan and Uzbekistan.

Kazakhstan consumed a total of 2.8 quadrillion thermal units of energy in 2012, with coal accounting for the largest share (63 percent). This was followed by oil and natural gas, at 18 and 16 percent, respectively.

Kazakhstan has successfully harnessed its oil resources to reduce poverty and boost shared prosperity.[1] The Kazakhstan national poverty line, based on income and minimum subsistence levels, dropped from over 44.5 percent of the population in 2002 to 2.8 percent in 2014 (figure 2.3). In the 2000s,

Figure 2.1 Production of Primary Energy Resources
Millions of tons oil equivalent

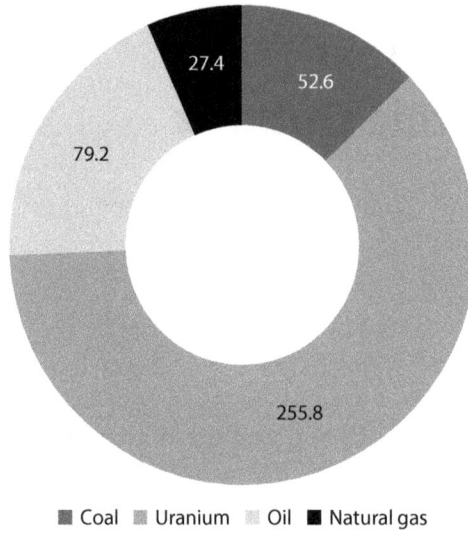

27.4
52.6
79.2
255.8

■ Coal ▨ Uranium ▧ Oil ■ Natural gas

Source: World Bank calculations based on data published by the Committee of Statistics of Kazakhstan.

Figure 2.2 Proven Reserves of Energy Resources
Billions of tons oil equivalent

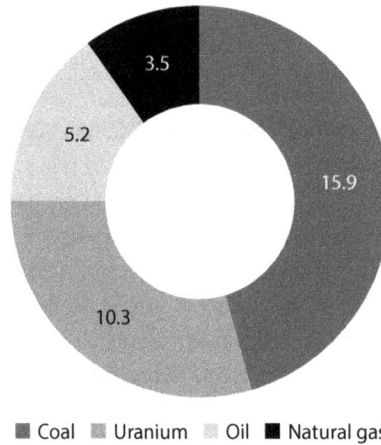

3.5
5.2
15.9
10.3

■ Coal ▨ Uranium ▧ Oil ■ Natural gas

Source: World Bank calculations based on data published by the Committee of Statistics of Kazakhstan.

poverty declined consistently. Growth in the income of the bottom 40 percent was systematically higher than growth of average gross domestic product (GDP) during this period, as reflected in the decline of the Gini coefficient from 0.28 in 2006 to 0.26 in 2013. Most of the poverty reduction in this period was driven by economic growth. Despite the progress in reducing overall

Figure 2.3 Poverty Rate in Kazakhstan, 2002–14

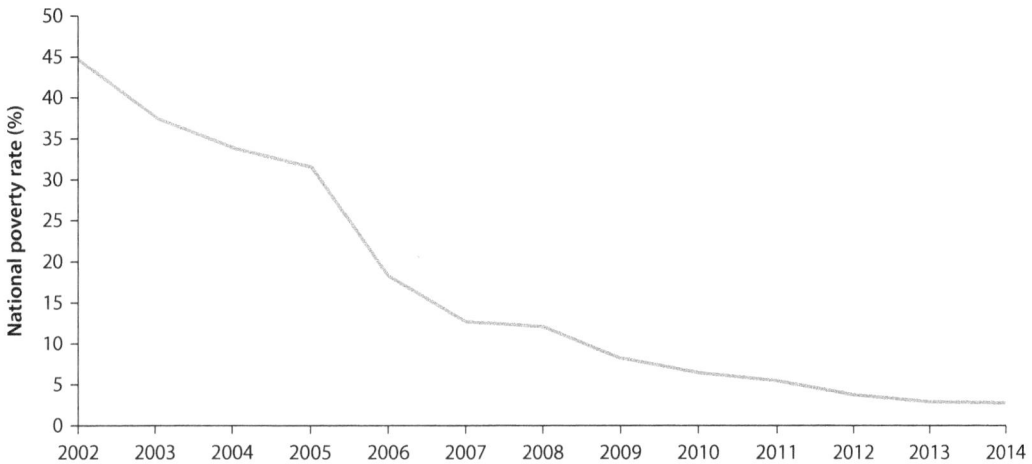

Source: World Bank calculations based on data published by the Kazakhstan National Statistics Office.

Figure 2.4 Poverty Rate, Middle Class, and Gini Coefficient, 2002 and 2014

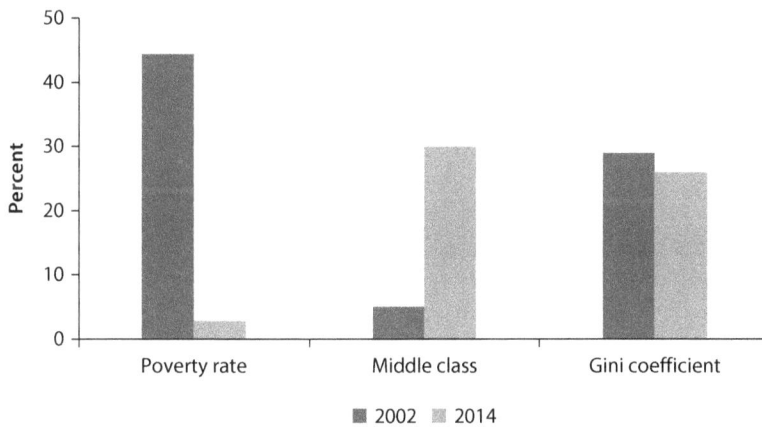

Source: World Bank calculations based on data published by the National Bank of Kazakhstan.

poverty levels, however, spatial disparities persist. The poorest regions, which are almost exclusively in rural areas, experienced smaller reductions in poverty during this period.

Kazakhstan faces a difficult external environment, resulting from the recent oil price decline and economic difficulties of Russia (its largest trading partner), as well as from the slower growth in China (another major market). This situation provides a window of opportunity to refocus the development agenda by strengthening the nonoil economy, by improving the private sector environment and reducing the current excessive role of the state in the economy.

Stuck in Transition • http://dx.doi.org/10.1596/978-1-4648-0971-2

The government of Kazakhstan has emphasized that reducing dependence on oil and facilitating the development of a well-functioning nonoil economy are high priorities. To this end, it has successfully managed its oil wealth and implemented prudent macro and fiscal policies.

Kazakhstan's economy bounced back from the global financial crisis of 2008/09, but growth slowed in 2014 because of weaker external demand and domestic imbalances. Kazakhstan's real GDP growth averaged 6.5 percent between 2010 and 2013, buoyed by higher oil prices. However, real GDP growth slowed to 4.3 percent in 2014 because of weaker domestic demand following the significant devaluation of the tenge in February 2014, the negative oil price shock during the second half of the year, and weaker external demand for exports.

Despite the fall in oil prices and weakened external position, the National Bank maintained the nominal exchange rate as a monetary anchor until mid-August 2015, at which point it shifted to a floating regime. The tenge immediately lost about 26 percent of its value (figure 2.5), falling from Kazakhstan tenge (KZT) 188.4/US$1 to KZT 254/US$1. The tenge continued to slide to about KZT 330/US$1 through April 2016.

A difficult external environment will continue to affect Kazakhstan's medium-term economic outlook. Over the longer term, diversification of the economy would increase its resilience to external shocks. Under the Base Case scenario presented in this study, oil prices are expected to remain low through 2016 before gradually recovering in 2017, supporting increased consumption and stronger investor confidence over the medium term. GDP growth will remain at about 1 percent in 2015 and 2016, but may increase to 3.3 percent in 2017.

Figure 2.5 Oil Prices and the Exchange Rate

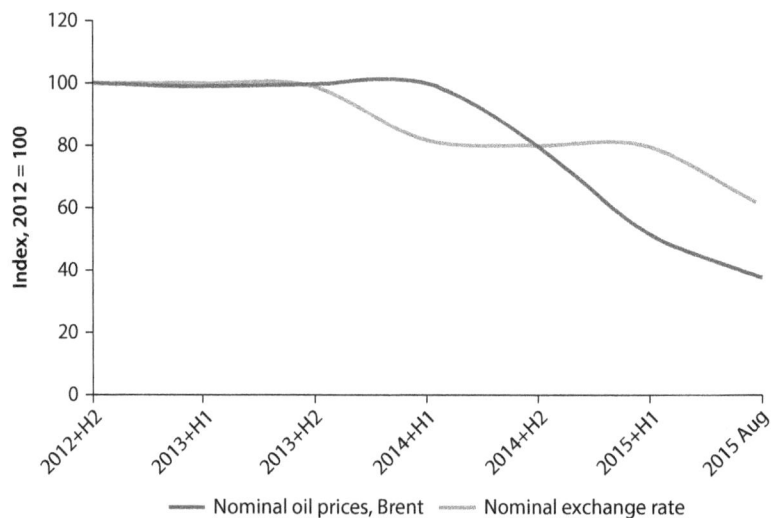

Nominal oil prices, Brent — Nominal exchange rate

Source: World Bank calculations based on data published by the National Bank of Kazakhstan.
Note: H1 = first half year; H2 = second half year.

The global oil supply is expected to outpace demand over the medium term, maintaining downward pressure on oil prices. The World Bank, under its baseline scenario, projected average oil prices (Brent-Dubai-West Texas Intermediate) at US$52.5 per barrel in 2015, US$51.4 per barrel in 2016, and US$54.6 per barrel in 2017 (figure 2.6).

In this context, it is assumed that Kazakhstan's GDP growth rate will be lower unless oil prices rise and output increases (figure 2.7). The uncertain external outlook will dampen private investment, and the pass-through effect of the tenge depreciation will reduce household consumption, with public consumption remaining modest because of the ongoing fiscal adjustment.

The rapid decrease in poverty has contributed to the fact that electricity remains relatively affordable for households. Energy averages about 5 percent of total consumption expenditures (figure 2.8), with those in poorer deciles spending proportionately more on energy than those in better-off deciles— although only marginally. Electricity expenditures are less than 2 percent of total consumption.

Expenditure on housing-related items is generally a small share of total consumption (figure 2.9). In the poorest decile, average household consumption of housing is KZT 15,465 per capita. For the richest 10 percent of the population, consumption in this category is higher, at KZT 71,056 per capita.

Energy consumption is a subset of the housing group as per the Classification of Individual Consumption According to Purpose. The poor spend less on energy than do the rich, and there is a difference in the type of energy consumed (figure 2.10). The poorest in the population are most likely to consume energy in the form of solid fuel. Among the rich, electricity is the most common type of fuel consumed.

Figure 2.6 Oil Output and Price Outlook

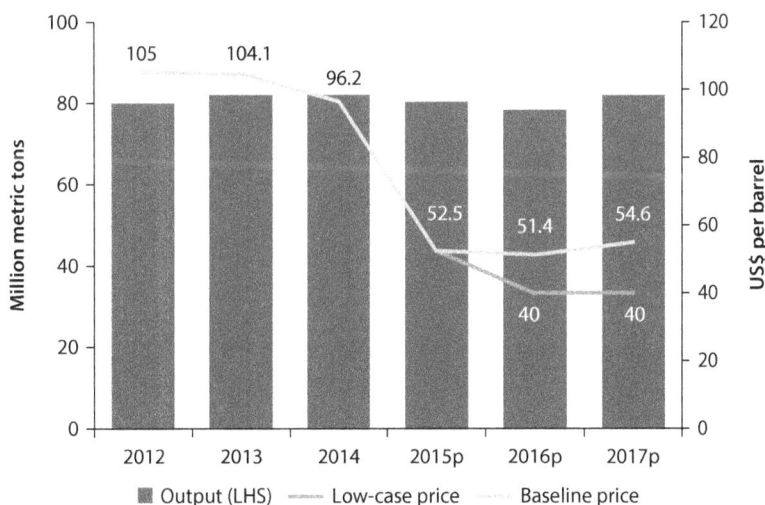

Figure 2.7 Kazakhstan's Gross Domestic Product Growth Outlook, 2012–17

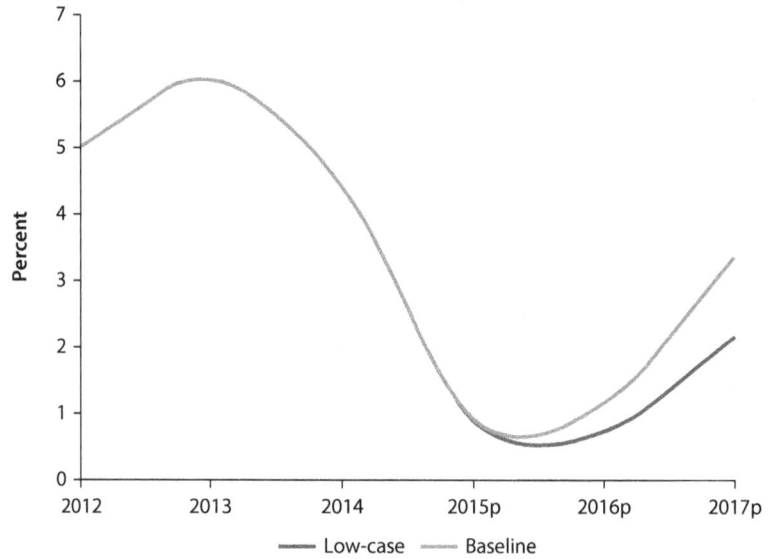

Source: World Bank calculations based on data published by the National Bank of Kazakhstan.
Note: The letter "p" represents prognosis.

Figure 2.8 Annual Consumption of Energy per Capita

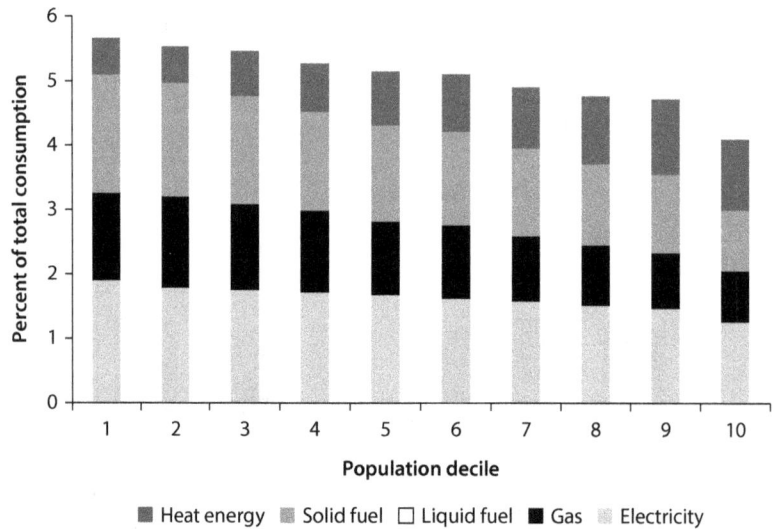

Note: Decile 1 = poorest 10 percent of the population; decile 10 = richest 10 percent.

Figure 2.9 Household Consumption Structure per Capita, by Decile

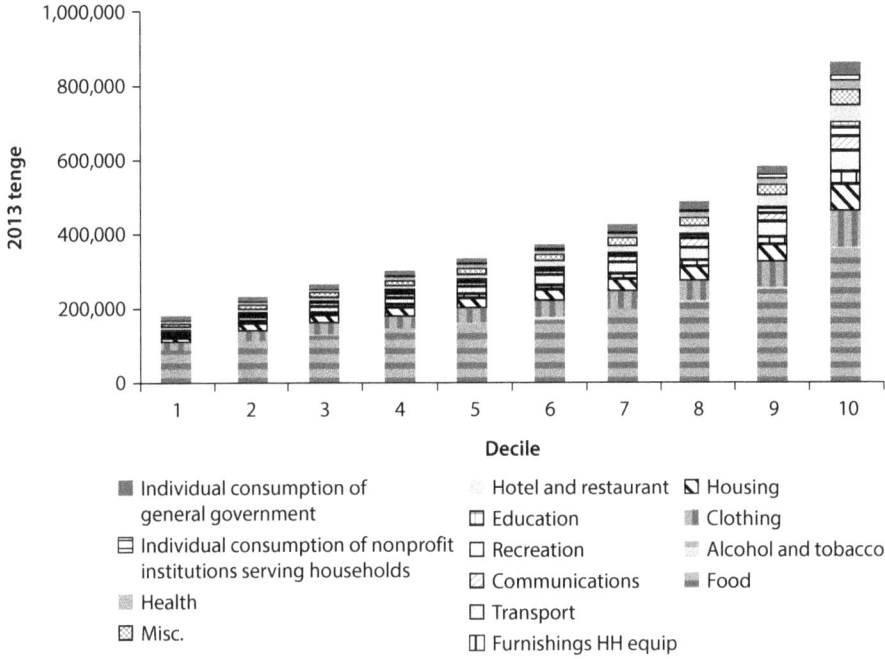

Individual consumption of general government
Individual consumption of nonprofit institutions serving households
Health
Misc.
Hotel and restaurant
Education
Recreation
Communications
Transport
Furnishings HH equip
Housing
Clothing
Alcohol and tobacco
Food

Source: World Bank calculations using the Kazakhstan Household Budget Survey.
Note: Decile 1 = poorest 10 percent of the population; decile 10 = richest 10 percent; HH = household. Energy is a subcategory of housing.

Figure 2.10 Annual Energy Consumption per Capita, by Decile

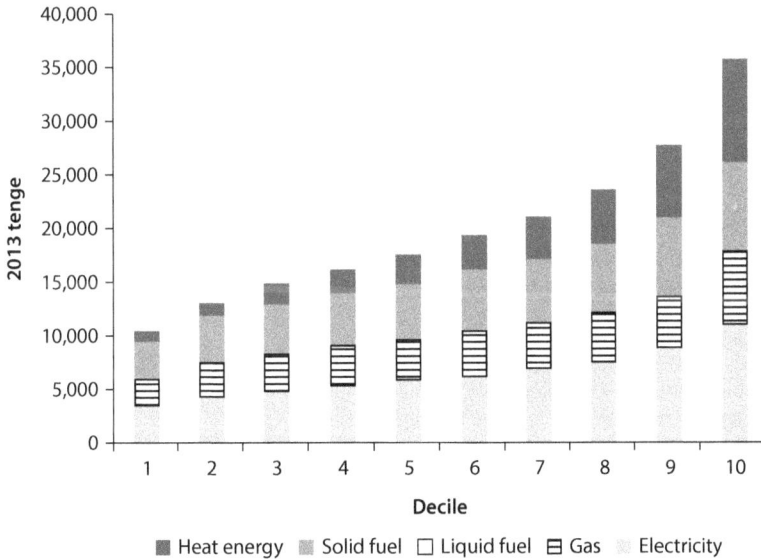

Heat energy   Solid fuel   Liquid fuel   Gas   Electricity

Source: World Bank calculations based on data published by the Committee of Statistics of Kazakhstan.
Note: Decile 1 = poorest 10 percent of the population; decile 10 = richest 10 percent.

Figure 2.11 Energy's Share of Annual Consumption per Capita, by Type and Decile

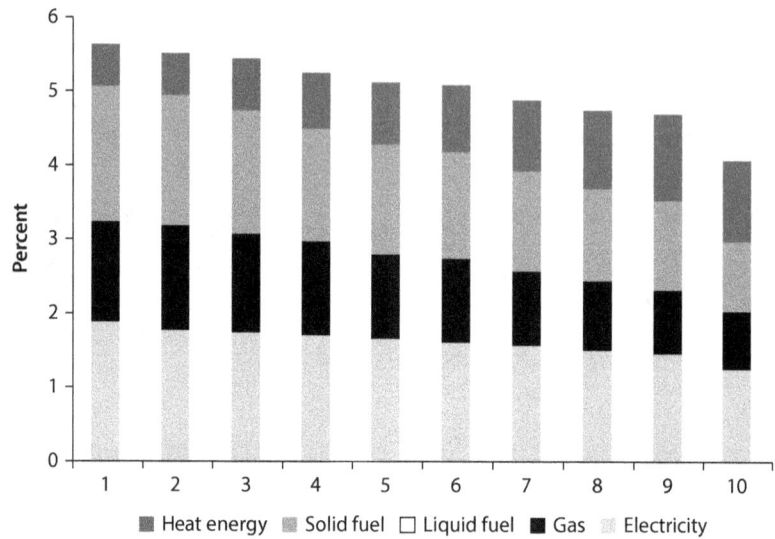

Source: Government of Kazakhstan 2014; Madani and Sarsenov 2015.
Note: Decile 1 = poorest 10 percent of the population; decile 10 = richest 10 percent.

The share of energy in total consumption exhibits variation across income deciles (figure 2.11). In the poorest decile, energy consumption comprises about 5.7 percent of total consumption, while it is 4.1 percent in the richest decile.

Note

1. From various sources including Madani and Sarsenov (2015) and Household Survey 2014, Statistics Agency, Kazakhstan (2014).

References

Government of Kazakhstan 2014. Household Survey 2014, Statistics Agency, Kazakhstan.

Madani, D., and I. Sarsenov. 2015. "Kazakhstan: Adjusting to Lower Oil Prices, Challenging Times Ahead—Biannual Economic Update (Fall)." World Bank, Washington, DC.

CHAPTER 3

Electricity Sector Overview and Status of Sector Reform

Overall Electricity Balance

More than 50 percent of Kazakhstan's primary energy is supplied by domestic coal, two-thirds of which is used in the power sector. More than 40 percent of the total primary energy supply is used to generate electricity and heat. Electricity and heat are about one-third of total final consumption. The energy balance shows that the ratio of total final consumption to total primary energy supply is less than 50 percent, compared with a world average of 69 percent (Sarbassov et al. 2013). The difference between total primary energy supply and total final consumption accounts for the energy used by traditional fuel supply sectors for extracting primary resources, transporting them, converting them to secondary fuels, and making them available to end users. In Kazakhstan this difference was equivalent to 35 million to 39 million tons of oil equivalent (mtoe) in 2007–10. The main components are (a) about 14–15 mtoe in thermodynamic transformation losses, mainly in power plants; (b) about 4–5 mtoe in distribution losses, mainly in district heating; and (c) about 15–20 mtoe in energy industry own uses, mainly in oil and gas extraction.

As of 2012, the total volume of proven primary energy reserves was 34.9 billion tons of oil equivalent. Coal, uranium, oil, and natural gas accounted for 45, 30, 15, and 10 percent of that amount, respectively. Coal-based generation accounts for 85 percent of electricity production,[1] followed by natural gas (7 percent) and hydropower (8 percent).

Around 72 percent of the country's coal is produced in the Pavlodar and Karaganda regions, located in the north central part of the country.[2] Coal reserves in the country are vast (34 billion tons) and many of them remain untapped. However, the great majority of coal produced is low-grade subbituminous and brown coal with relatively low heating value and large ash and moisture content.[3] The majority of coal mining is done in the eastern and northern parts of the country, and is practically absent in the western and southern regions.

The largest share of power production in the Western and Southern zones is from natural gas. As of 2010, the proven reserves of natural gas were estimated at around 3.9 trillion cubic meters (m^3).[4] The electricity-isolated Western zone is run entirely on natural gas. Almost all remaining natural gas reserves are located in sub-salt deposits of the Peri-Caspian Lowland on the coast of the Caspian Sea. The great majority of natural gas is associated with oil extraction. Forty-six percent of natural gas production (40 million m^3 in 2012) is reinjected into the oil formation, while a large portion of the remainder is used to satisfy oil extraction needs (with power produced by gas turbines). A portion of the turbine output is directed to the power grid. In 2012, 11 million m^3 of refined market-able gas were produced, 46 percent of which was consumed by the power indus-try. Currently, only the Southern and Western zones have sufficient access to the domestic natural gas transport system, which is used mainly for the transit of Russian and Central Asian natural gas to third countries (including the European Union and China), and to transport Kazakhstani gas to be sold to the Russian Federation (map 3.1).[5]

Gas processing facilities are also located in the Southern and Western zones. As a result, this region with low potential demand is supplied with Kazakhstani, Russian, and Turkmen gas. The northwest area—comprising the Aktobe and

Map 3.1 Main Trunk Natural Gas Pipelines

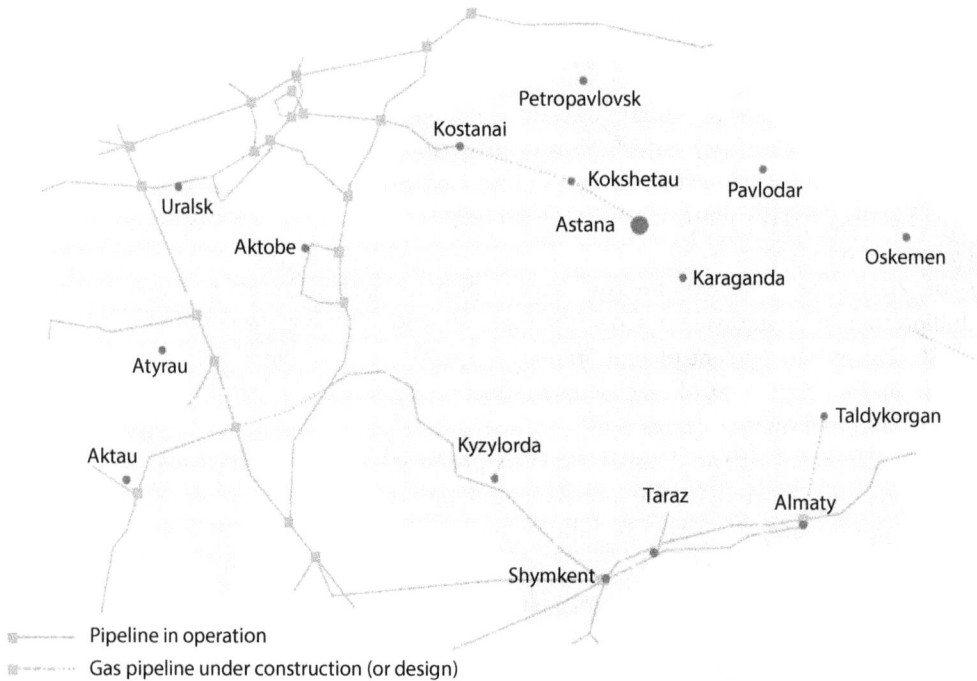

⊞------ Pipeline in operation
⊞······· Gas pipeline under construction (or design)

Source: KazEnergy 2013.

Kostanay regions—is supplied with gas through the Bukhara-Ural gas pipeline and from the Zhanazhol field. The southernmost populated and industrially developed area (Almaty, Zhambyl, and the Southern Kazakhstan regions) is interruptedly supplied with insufficient volumes of gas from Uzbekistan through the Bukhara Gas Region-Tashkent-Bishkek-Almaty gas pipeline (KazEnergy 2013).

In summary, (a) coal mining and consumption are limited to the northern and eastern parts of the country; (b) natural gas production and consumption are concentrated in the western and southern parts of the country; and (c) natural gas reserves are large enough to supply the country's entire energy needs for many decades. Thus, a transition to a natural gas economy—mainly for environmental reasons—is possible if the gas transport infrastructure becomes physically available across the country. Although such a plan exists for extension of the current infrastructure, the proposed additional 67,000 kilometers of pipelines have low financial attractiveness at this time.

Kazakhstan is second in the world in uranium reserves (629,000 tons of recoverable uranium reserves in 2012) and first in production. The country has had experience with uranium enrichment since the former Soviet period. In addition, Kazakhstan has experience operating the BN-350 power reactor, the world's first industrial fast neutron reactor, which was in operation from 1972 to 1999. Rebuilding the capacity to enrich fuel and constructing nuclear power capacity may be a strategic choice for the country to reduce carbon dioxide emissions. Currently, there are plans for the construction of nuclear capacity of up to 1,000 megawatts (MW) in the eastern part of the country.

Power Sector Institutions

Kazakhstan's Ministry of Energy is a policy-setting institution that oversees the electricity sector. In August 2014, the President of Kazakhstan announced an extensive government reorganization with the intention of creating a more compact and effective government. The number of ministries was reduced from 17 to 12, and a unified Ministry of Energy was created, absorbing the functions of the Ministry of Oil and Gas and parts of the functions of the Ministry for Industry and New Technologies and the Ministry for Environment and Water Resources. Technical regulation is carried out by the Committee of Atomic and Energy Control and Supervision under the Ministry of Energy. Figures 3.1 and 3.2 and map 3.2 depict various aspects of the institutional structures in place for the power sector.

Tariff regulation is carried out by a Competition Protection Committee under the Ministry of National Economy. The state-controlled regulatory system has evolved in the past two decades, although the sector regulation still lacks an adequate degree of autonomy and is vulnerable to political interference (at the national and regional levels) and varying degrees of "regulatory capture" by powerful incumbent sector entities. The governance structure (including terms of appointment of key regulatory officials and source of budget) is not in line with best practice. Overall, the existing regulatory system is a major shortcoming to

Figure 3.1 Kazakhstan's Power Sector: Institutional Structure

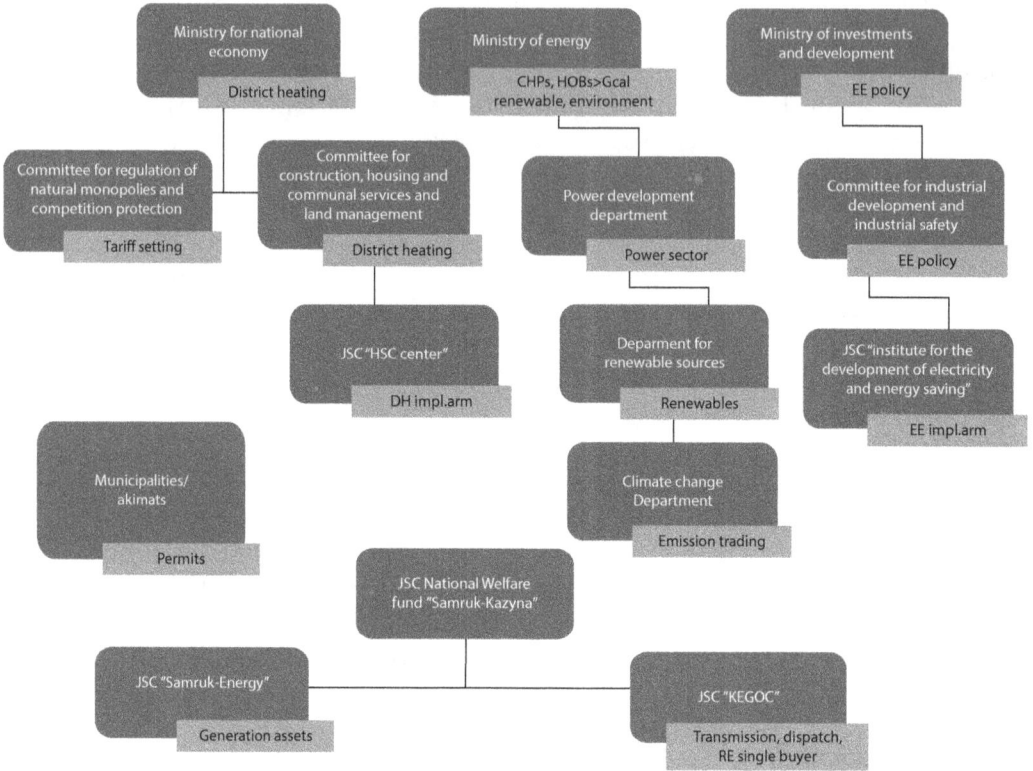

Note: CHPs = combined heat and power plants; DH = district heating; EE = energy efficiency; GCal = gigacalories; HOB = heat-only boiler; HSC = Housing Services Center; JSC = Joint Stock Company; KEGOC = Kazakhstan Electricity Grid Operating Company; RE = renewable energy.

Figure 3.2 Kazakhstan's Power Sector Unbundled into Generation, Transmission, and Distribution Subsectors

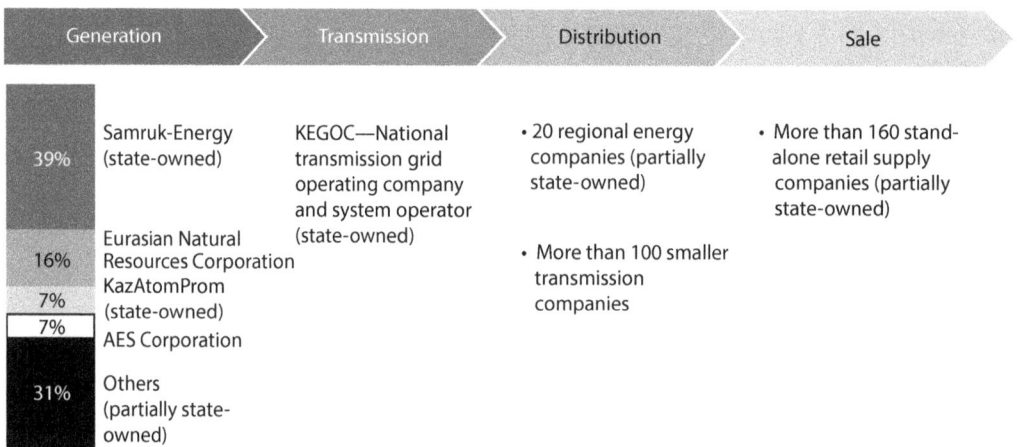

Generation	Transmission	Distribution	Sale
39% Samruk-Energy (state-owned)	KEGOC—National transmission grid operating company and system operator (state-owned)	• 20 regional energy companies (partially state-owned)	• More than 160 stand-alone retail supply companies (partially state-owned)
16% Eurasian Natural Resources Corporation		• More than 100 smaller transmission companies	
7% KazAtomProm (state-owned)			
7% AES Corporation			
31% Others (partially state-owned)			

Note: KEGOC = Kazakhstan Electricity Grid Operating Company.

Map 3.2 Regional Zones of the National Power System

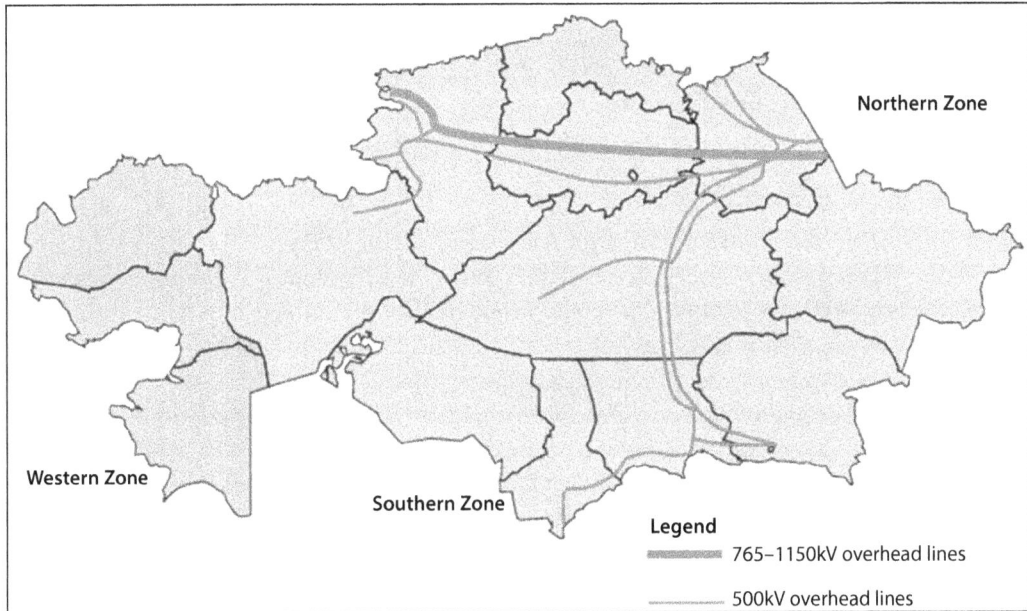

Note: kV = kilovolt

be addressed by the government. The new energy strategy (Energy Concept 2030) devotes only scant attention to regulation. A professionally managed and reasonably autonomous regulatory institution is crucial to (a) improving the performance of the natural monopoly segments of the power sector, (b) supporting competition in the contestable market segments, and (c) attracting investors to the sector. International evidence shows that a well-designed, credible regulatory system reduces the cost of private capital for the power sector. This is an important benefit to reap in Kazakhstan's current capacity-constrained system. The recent incorporation of the former stand-alone Agency for Regulation of Natural Monopolies (AREM) into a line ministry is a step backward on the road toward more autonomous and depoliticized sector regulation that constitutes an essential element of an attractive investment climate in the power sector.

The government of Kazakhstan created the National Wealth Fund, Samruk-Kazyna, in October 2008, merging the State Assets Management Holding Company, Samruk, and the State Sustainable Development Fund, Kazyna. The consolidated institution is tasked with ensuring efficient management of state assets, including in the power sector. As such, the institution owns and controls Samruk-Energy, the largest electricity generator, and Kazakhstan Electricity Grid Operating Company (KEGOC), the national high-voltage-transmission company and system operator. In addition, the Kazakhstan Operator of Electric Energy and Capacity—the spot market operator—is also a subsidiary of the Samruk-Kazyna holding.

The national power system is split into three regional areas (referred to as operating zones): Northern, Southern, and Western, with the latter isolated and disconnected from the rest of the national system (map 3.2).

Corporate Governance

As a result of the recent renationalization of considerable generation assets, the state's role in the power sector, again, is very strong. Improving the corporate governance[6] of state-owned companies across the electricity value chain is a major task, but it has not yet received adequate attention from the government. This is a largely neglected aspect in the Energy Concept 2030, despite the indication of considerable room for improvement across key corporate governance issues. Examples of room for improvement include development of an effective legal and regulatory framework, ownership policies that incorporate transparency and accountability, and equitable treatment of and relationship with shareholders. Considerable cross-country evidence shows that the costs of not reforming state enterprises are high, and sustained government efforts are needed to improve their performance. This can be done by improving privatization policies and exposing state enterprises to market discipline through the new private entry and exit of unviable firms and improvements in their management (Kikeri and Kolo 2005).

Good corporate governance, including transparency, is another essential component of a country's investment climate. Research shows that investors are willing to pay a premium of 10–12 percent for the shares of companies having a credible corporate governance framework in place (McKinsey & Company 2002). In this context, the government's newly launched privatization drive in the energy sector may enhance the quality of corporate governance. The regulatory framework (overpoliticized and lacking sufficient autonomy) and monitoring and control (both weak) of market power are in need of considerable strengthening. Corruption is another important corporate governance issue to address, all the more so because indications suggest considerable shortcomings in this area.

Generation

Electricity in Kazakhstan is generated by more than 100 power plants with a total installed capacity of 20.8 gigawatts (GW) and available capacity of 15.2 GW at the end of 2014 (available capacity is derated thermal capacity + 30 percent firm hydro capacity). Kazakhstan's total net generation in 2015 was 90.8 billion kilowatt hours (kWh) of electricity, with over 81.6 percent coming from coal-based thermal power plants (TPPs), 8.0 percent from gas-fired plants, 10.2 percent from hydropower plants (HPPs), and less than 1 percent from renewable energy (including small hydro). In 2013, about 70 percent of the installed capacity was considered technically obsolete. The average age of the TPPs was 28.8 years (with 57 percent of them in operation for 36 years). Their thermal efficiency is 32 percent on average compared with 42 percent in the leading Western power systems. Fifty-seven percent of the HPPs are more than 30 years old. The environmental performance of most TPPs is poor, despite the fact that 45 percent of atmospheric pollution comes from the electricity sector (KazEnergy 2015).

Most of Kazakhstan's power generation comes from coal-fired power plants, which are concentrated in the north of the country near the coal mines (figures 3.3 and 3.4). The largest TPPs include

- Ekibastuz Regional State Power Station (GRES) GRES-1,
- Ekibastuz GRES-2,
- EEC Corporation Power Plant,
- Kazakhmys Energy GRES, and
- Zhambyl GRES.

The largest HPPs are primarily used to meet peak demand and include

- Bukhtarma HPP,
- Ust-Kamenogorsk HPP, and
- Shulbinsk HPP.

The largest combined heat and power plants (CHPs) include

- Karaganda CHPP-3 (by Energocenter),
- Temirtau CHPP-PVS, CHPP-2 (by Arcelor Mittal),
- SSGPO CHPP (by ENRC Corporation),
- Balkhash CHPP and Zhezkazgan CHPP (by Kazakhmys Energy), and
- CHPP-1 (by Aluminum of Kazakhstan and ENRC Corporation).

Figure 3.3 Installed Capacity, by Fuel Type and Zone

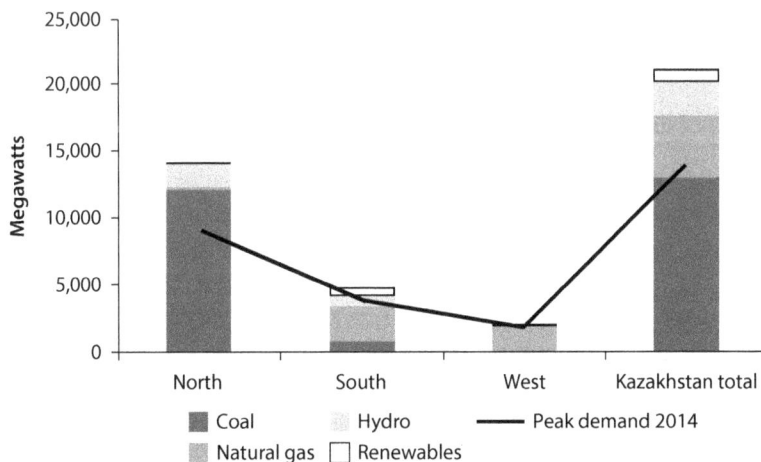

Figure 3.4 Derated Capacity, by Fuel Type and Zone

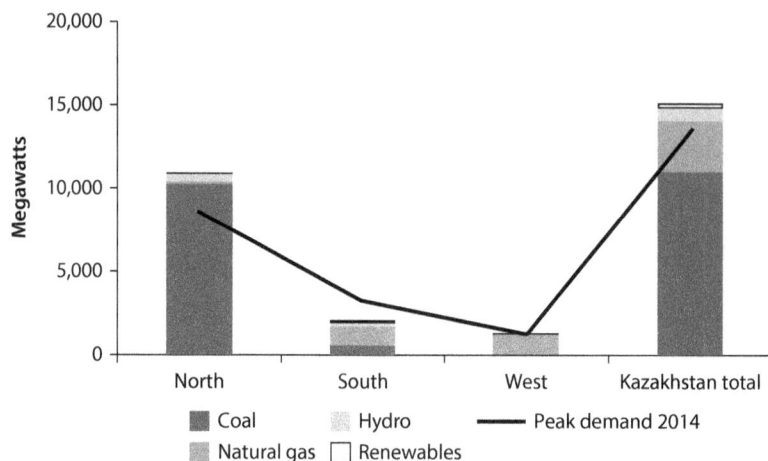

Given Kazakhstan's significant uranium deposits, plans have long existed to build additional nuclear power plants, although little progress has been made to date. The government is considering the construction of two nuclear power plants near the town of Kurchatov in the northeast part of the country and on the northern shore of Lake Balkash, in the south.

Ministry of Energy data show around 17.1 GW of installed thermal capacity in 34 TPPs, but the available capacity is 14.1 GW. The installed hydropower capacity is 2.2 GW, with an average capacity factor of 40 percent.

Following independence, almost all generators were privatized to local and foreign strategic investors—several under concession agreements. However, because of the departure of several foreign investors, private ownership fell considerably. The creeping renationalization process has accelerated recently under the vigorous acquisition strategy of Samruk-Energy, a state-owned holding company under Samruk-Kazyna. Presently, Samruk-Energy owns about 40 percent of the country's generation capacity, which poses a risk to the preservation of competitive conditions in the subsector. The bulk wholesale market has rapidly evolved toward an oligopolistic structure.

In 2014, the government initiated the "second wave of privatization." Under this program, only two generation subsidiaries of Samruk-Energy were planned to be wholly or partially divested to private entities. The fact that Samruk-Energy and KEGOC are owned by the same parent company (Samruk-Kazyna) raises the risk of excessive market power at the wholesale level. This concern is legitimate considering the tight capacity situation prevailing recently. The competition problem is further compounded by the fact that several power plants are owned and operated by large industrial enterprises for self-generation,[7] with much of the power produced not entering the national wholesale market, thereby limiting the scope of competition among generators.

Most generators, even the smaller CHPs, are considered "dominant" at the national or regional level in accordance with antimonopoly legislation, and are listed in the respective registry of dominant firms. However, enforcement of the antimonopoly legislation is lax.

In addition to hampering competition, the potential danger of a highly concentrated generation market is that firms with market power may not have sufficient incentives to invest. Indeed, withholding new investment could be a means for dominant firms to push up prices and increase profits (IEA 2002). For a considerable period of time after independence, the large number of generators, many of them privately owned, and the existence of a massive generation overcapacity inherited from the former Soviet Union created fertile ground for intense "cut-throat" competition among generators for cash-paying customers in the liberalized wholesale market. As a result, and because of the extremely low coal prices, generation tariffs were depressed and barely covered the operating costs of generation. This situation left few resources for long-term modernization and asset replacement, let alone systematic capacity expansion to prepare the sector for the coming sharp upswing in power consumption from 2010 onward. For a long time, the generation side remained stuck in short-run survival mode. The government's policies were unable to strike the right balance between complete liberalization and adequate resources for generators to modernize and undertake expansion. Further, the excess capacity obscured the need to think long term. Amid the highly unpredictable and unsophisticated regulatory and investment climate, the "energy-only" market prices were too low to attract serious industry players willing to invest in generation.[8] Investments were insufficient even for standard maintenance and rehabilitation, let alone increased capacity.

Toward the end of the 2000s, the vigorous growth in electricity demand and lack of large-scale investments in new generation led to the virtual disappearance of the country's large surplus generation capacity. Kazakhstan faced an impending capacity shortage with the associated risk of an economywide electricity shortage. Such a shortage could drag down the economy and trigger a tariff hike, and fuel inflation and thereby threaten social stability and reduce the export competitiveness of energy-intensive exports in the external markets. In an inhospitable regulatory environment, the power sector faced a deep challenge in attracting new investors.

The generation capacity margin rapidly and steadily shrank from 53 percent in 2000 to a dangerously low 4 percent in 2012 (figure 3.5).[9] Especially amid the winter peak load, the Kazakhstan electricity system faces an uncomfortably slim safety cushion of reserve power capacity and the risk of unscheduled blackouts in the absence of interruptible contracts with large consumers, whereby the latter would be willing to reduce their consumption at critical times of low spare capacity.[10] The considerable risk of plant breakdowns or forced outages, caused by technical obsolescence and poor maintenance of much of the power fleet, further enlarged the potential for emergency measures, spot shortages, and blackouts. Most power plants were grossly inefficient,[11] and a significant part of the

Figure 3.5 Tightening Supply and Demand Balance and Reserve Capacity, 1993–2014

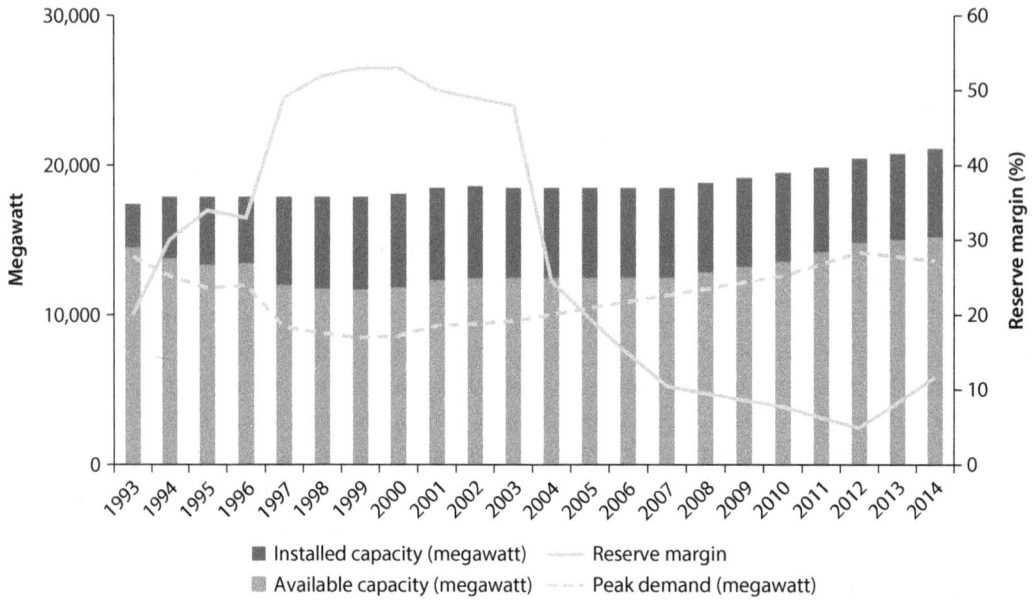

Legend:
■ Installed capacity (megawatt)   Reserve margin
■ Available capacity (megawatt)   - - - Peak demand (megawatt)

installed capacity was unavailable for operation. In addition, the environmental performance of the plants fell short of accepted international norms by a large margin. Compounding matters further, Uzbekistan's repeated, large-scale unscheduled power "imports" from Kazakhstan—within the framework of the synchronously interconnected Central Asian Power System—further strained the already tight reserve cushion.

Amid alarming prospects of a debilitating capacity crunch, the government faced two major choices to tackle the long-term reliability and security challenge of the power system. One was to proceed with the aggressive market liberalization that started in the mid-1990s, and allow electricity prices to rise to reflect the underlying supply and demand gap. This approach would provide appropriate signals for the market to attract investors in the hope of high prices. These were normal market pressures created by the underlying demand and supply fundamentals. However, the government considered that it would be too risky to rely completely on a pure market mechanism in a situation that was increasingly perceived as a national economic emergency that required quick results that the market would be too slow to deliver. There was also a strong sense that the power sector offered fertile ground for speculators seeking short-term opportunities to "loot" the system and take advantage of a tightening supply and demand balance. It was claimed that some plant owners would siphon off some or all of the revenues earned from higher prices instead of investing in the sector.

Because of these concerns, policy makers wanted to create a system—an explicit compact with generators—that would simultaneously coerce and incentivize all major generators to make the required investments. This would be in parallel to ensuring that increased revenues from tariff increases would be used solely for the badly needed investments in additional reliability and capacity instead of flowing out of the sector (KEMA 2013). Under this second option— and given the absence of an attractive investment environment and an organized capacity market—the government hurriedly resorted to two approaches.

The first approach was the ad hoc "Balkhash" investor-attraction model, whereby under a long-term power purchase agreement (PPA), Korean investors were awarded a license—without competitive bidding—to build a 1,320 MW coal-fired power plant along Balkhash Lake. Although labeled as an independent power producer model, this approach has clearly been unsustainable for the long term. Incidentally, this model has not worked well even for the Balkhash plant, per se.[12]

The second (mainstream) approach involved the government's introduction of an administrative generation tariff regulation in 2009. Under this system, tariff caps (also referred to as maximum or "investment" tariffs) were imposed on all major generators (including private ones) to make new investments in modernizing and extending capacity. It is a state-managed investment commitment scheme: "higher tariffs for new investments." It also includes severe restrictions on the use of profits resulting from higher generation tariffs.

Under a complicated and not fully transparent scheme, generators were grouped into 13 "tariff groups" on the basis of plant type, fuel used, and distance from the fuel source. Within each tariff group, an escalating upper limit (cap) was set for a seven-year period, and has been adjusted annually. All generators have been legally mandated to develop a medium-term investment program. Each generator has undertaken a specific investment program in return for a tariff increase, which is not to exceed the cap for the given tariff group. As expected, the actual tariffs have moved relatively close to the caps. For example, for the large regional state power station, Ekibastuz GRES-1, the tariff cap and the actual tariff were KZT 8.0/kWh (or 4.4 U.S. cents/kWh) and KZT 7.12/kWh (4.0 U.S. cents 4.0/kWh), respectively, in 2014. Penalties are imposed for not meeting the investment obligations. Pursuant to the Electricity Law, this system was envisaged to remain in place until the end of 2015.

The allowed tariff increases have been substantial. For example, in Category 1—which comprises three major coal-fired power plants accounting for the bulk of the Ekibastuz-based power generation—tariffs have been allowed to rise nearly 2.5 times (or 25 percent per year) between 2008 and 2015 (see appendix A). The tariff hikes allowed generators to fund nearly half of their investments from own resources; the balance was financed by the government and, to a small extent, by commercial loans.[13] As expected, the actual tariffs were relatively close to the allowed maximum tariffs.

To its credit, the government program produced a mini-investment boom of 28 percent per year on average between 2009 and 2015—a steep increase over the previous period. Investments undertaken between 2009 and 2014 amounted to KZT 2,230 billion (about US$14 billion at the average exchange rate) and resulted in rehabilitation of about 5,000 MW of existing capacity and an additional 1,700 MW of new capacity.

The rehabilitated and expanded generation capacity improved system reliability. For example, the number of emergency outages at the major power stations of national importance fell from 131 in 2008 to 39 in 2013, thereby reducing the hours of outages from 3,200 to 900 per year. The gap between installed and available capacity was narrowed by 20 percent. The earlier generation capacity deficit was eliminated and, in 2014, the Kazakhstani system had a surplus of about 1,600 MW of available capacity, which translates to a reserve margin of 12 percent—a significant improvement over the slim reserve margins previously prevailing. Falling demand growth in the wake of the recent oil price–induced economic recession has further eased pressure on the generation sector.

Legitimate concerns have arisen, however, about the efficiency, transparency, and long-term sustainability of the investment tariff program. Although substantial new investments have indeed been undertaken, the program is still an inefficient, nontransparent scheme, open to much bargaining and administrative discretion. Some Kazakhstani sector officials themselves and independent experts consider the program inefficient and too expensive, and the large tariff increases unsubstantiated. A government audit of the program, covering 2010–14, found several shortcomings, including "inappropriate utilization" of the incremental tariff revenue, meaning that it was not fully utilized for capacity modernization and expansion. In some cases, a part of the revenue was used for financial investments or for paying higher dividends to the owners. Indeed, the caps led to significant increases in generation tariffs and much-improved finances for some of the generation companies in several "tariff groups," although their levels are still too low to make large-scale greenfield generation projects commercially viable. Clearly, the tariff caps arbitrarily chop off some appropriately high prices (that is, prices that correctly reflect the underlying capacity shortage). To that degree, the price caps reduce the revenue stream for generators, thus creating a shortfall in total returns to their investments. The existing administrative tariff framework is unattractive to foreign investors.

Concerns also arise on whether the investments carried out were economically efficient (that is, achieved at least cost or at least at a reasonable cost). For 2009–14, the combined rehabilitated and expanded capacities were rather expensive, at an average US$2,100 per kilowatt (kW). This is comparable to capital costs for new coal-fired generation. However, only one-fourth of the combined rehabilitated/expanded capacity was new.

The return to generation tariff regulation by way of price caps in the competitive segment of the electricity industry was a major step backward from the earlier competitive evolution of the Kazakhstani power market, amounting to government micromanagement of the investment process. It is unlikely that the maximum tariffs and the investment commitments imposed on generators were at economically efficient levels. The tariff caps for each generator group are a government estimate, made by an administrative process of the required or acceptable levels of maintenance, rehabilitation, modernization, or new construction for each group. The process involves heavy two-way bargaining. Power plant owners and operators are better placed to make investment decisions without having to justify them in a bargain-filled bureaucratic process. Because of these shortcomings and the fundamental inconsistency of the process with the underlying principle of a liberalized power market—the professed goal of the government's new sector strategy under the Energy Concept 2030—the tariff cap system does not provide an efficient solution for Kazakhstan's longer-term investment requirements. Nevertheless, the government decided in 2015 to extend the tariff cap for an additional seven-year period (2016–22), while increasing the number of tariff groups to 16 and freezing tariffs across the new period at the level set for 2016.[14] As a novel feature of the program, in addition to the energy tariffs, capacity availability tariffs were set (in tenge/MW/month) for the same period, freezing them at the 2016 level. Therefore, from 2016, a two-tier administrative tariff regime will be in place with a capacity charge component while the government sets cap tariffs for energy for a medium-term period. Clearly, in the high inflation environment facing Kazakhstan following the large devaluation of its national currency since August 2015, the tariff freeze will not be sustainable, and a return to annual revisions is very likely. In conjunction with the prolongation of tariff caps, introduction of the capacity market was delayed from January 1, 2016, to January 1, 2018.

There is a daunting investment mobilization challenge to implement the government's ambitious generation expansion plan under the Energy Concept 2030: 7,500 MW of new capacity in the period 2013–30, estimated to cost about US$5.5 billion (US$325 million per year). This study (chapter 6) projects— under the Base Case scenario—an even higher annual average investment need of US$820 million over 2015–45. The currently envisaged and controversial generation market arrangements (price caps and a much delayed capacity market) may well prove insufficient to meet the investment challenge, let alone in a least-cost manner and relying mostly on private sector financing.

A key obstacle to greater private sector participation is the highly uncertain investment climate. The government is continuously changing investment laws— it amended them five times in 2011, 39 times in 2012, and 270 times between 2014 and 2015, making the investment environment highly unpredictable. The government hopes that Kazakhstan's accession to the World Trade Organization in December 2015 will motivate investors to look more positively on the country's power sector.

The government's new "second wave privatization" program for the generation sector is too limited to reverse the recent renationalization trend. In addition to political will, the much-needed, large-scale privatization hinges on a substantial overhaul of the sector's legal and regulatory framework.

Transmission

Kazakhstan's national transmission grid (220 kilovolts and higher) is operated by KEGOC, a state-owned company and subsidiary of Samruk-Kazyna (map 3.3). KEGOC is responsible for electricity transmission and network management.

Map 3.3 Kazakhstan's National Transmission Grid

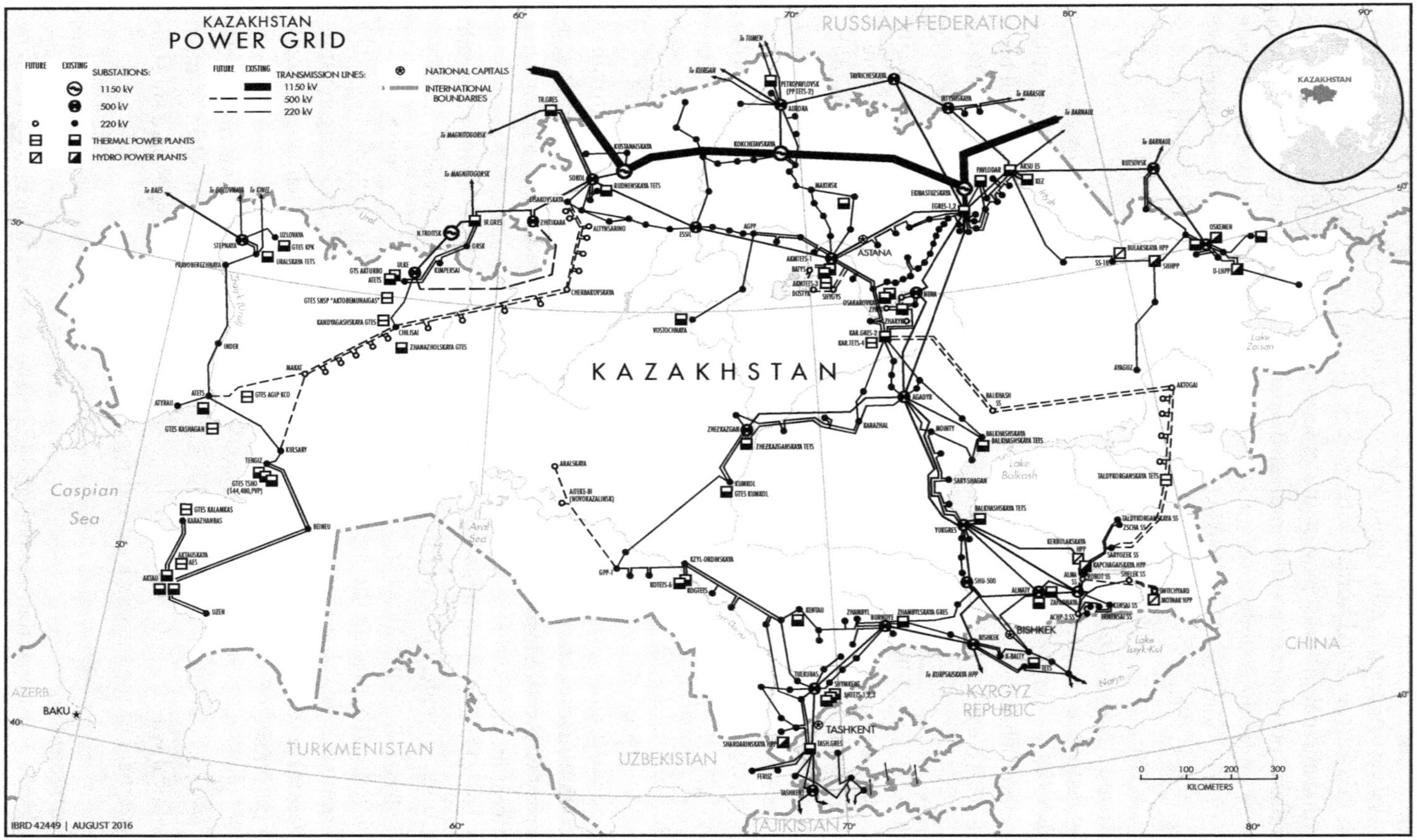

Source: Kazakhstan Electricity Grid Operating Company.

In addition, as the system operator, KEGOC is responsible for central dispatch control, system security, and international connections. As a result of large-scale rehabilitation and extension investments, the reliability performance of the transmission system has improved considerably.[15] Subject to system security and network constraint, the existing system of central dispatch is, in effect, a self-dispatch by generators against bilateral contracts with customers.

Nondiscriminatory access rights of third parties to the high-voltage transmission grid are a sine qua non for a competitive power market and increased electricity trade. Government legislation (the Electricity Act and Grid Code) ensures equal access to the transmission grid (national or interregional) to all qualified wholesale market participants. However, the access regime is unsophisticated, lacking clear and detailed procedures, protocols, transparency, and a credible dispute resolution mechanism. This circumstance contributes to the poor investment climate as generally perceived by potential investors in the Kazakhstani power sector. The market uncertainties confronting a new generator include the future demand and price, although, in addition, the generator must have assurance of open, nondiscriminatory access to adequate transmission capacity to deliver power to the purchaser. The probability that transmission capacity might be constrained (including arbitrarily by the grid operator), with the effect of being denied access to the grid, constitutes an additional element of uncertainty to the prospective generator. Such constraints may arise for at least two reasons—congestion and strategic behavior.[16] Over the past two decades, portions of the Kazakhstani transmission grid—in particular, the North–South interconnector—were often congested. Such congestion is evidenced by a significant number of transmission load reliefs. Faced with uncertainty about future transmission availability, a new generator will factor into its revenue calculation some risk that output may not necessarily find its best market. Therefore, investment in such a plant will require compensation for the risk posed by transmission constraints.

Another element of transmission uncertainty arises as a result of a great deal of the transmission grid in Kazakhstan being owned by vertically integrated utilities.[17] Despite open access requirements, allegations of preferences for own generation and limitations on carrying power for other suppliers continue. The effect of such actions is to raise further the probability of a lack of transmission or its availability on nondiscriminatory grounds. This, too, constitutes an impediment to new entry.

Against this background, the existing system can be best described as "minimal open access," which involves more discretionary intervention and less based on market rules. Because of locational congestion constraints, KEGOC is occasionally left with considerable discretion as to which sale/purchase transaction to execute and which to refuse to schedule. In the absence of published protocols covering such contingencies, KEGOC's discretion is significant. In a more sophisticated open access regime—where, for example, there are nodal prices and transparency about real-time availability of transmission capacity—the question of which competing transactions should go forward under congestion would be resolved by the parties themselves, one of whom agrees to pay the congestion rent. The market, not KEGOC, would decide which transactions take priority in the queue.

Effective locational price signals are particularly important for building new transmission capacity. Currently, in the absence of meaningful price signals, the costs of grid congestion are inefficiently "socialized"—by applying the so-called "postage stamp" tariff methodology—across the entire power system. The practice of socializing the congestion and incremental costs of transmission discriminates in favor of electricity resources distant from load centers, such as generation in the northern Ekibastuz hub; and it fails to reward appropriately the strategic location of new generators closer to new load centers.

Network congestion management is a major function of the system operator to ensure that the transmission system does not violate its operating (security) limits. Congestion management is extremely important in a large and bottle-necked network such as Kazakhstan's; if it is not properly implemented, it can impose a barrier to trading in electricity. Even with the recent substantial expansion of the high-voltage grid—mostly in the north–south direction—network constraints still plague the Kazakhstani system from time to time, particularly in severe peak load situations.[18]

Ideally, an effective congestion management system enables all network users to compete for scarce transmission capacity on a level playing field. Historically, KEGOC has employed a nontransparent quota system to allocate the congested network capacity—while balancing the system in real time—occasionally making itself vulnerable to charges by market participants of biased, non-independent management of network constraints.[19]

Another area of concern is the provision of ancillary services, such as voltage control, reactive power, black start, and so on. In a "minimal open access regime," such services are provided on the basis of arrangements by the system operator, which is typically nontransparent, and are bundled into a generic charge imposed on all users of the grid. Currently, the arrangements in place and the lack of transparency greatly limit the ability of generators to engage fully in providing such services. This means that the transmission grid is open to some transactions—mostly standard sell/buy deals—but closed to others.

Distribution and Retail Sale

Regional grids (up to 220 kilovolts) are operated by 20 regional electricity distribution companies (RECs), which also own generators (mostly CHPs) to carry out transmission, distribution, and supply (by their subsidiaries) of electricity within their service territories (map 3.4). Some of them were privatized, while others are municipally owned or belong to Samruk-Energy. Because of the unpredictable regulatory framework, which some market players see as intrusive, several high-profile foreign investors departed from the distribution subsector.

The distribution system is excessively fragmented. For example, the Akmola, Almaty, and Karaganda regions have more than one REC, often under nontransparent ownership. Furthermore, in some regions, many small distribution companies directly interface with retail customers. For example, as many as 42 distribution entities operate in the Karaganda region. Most of the distributors are in a difficult situation, technologically and financially. In Kazakhstan, no

Map 3.4 Location of Regional Electricity Distribution Companies

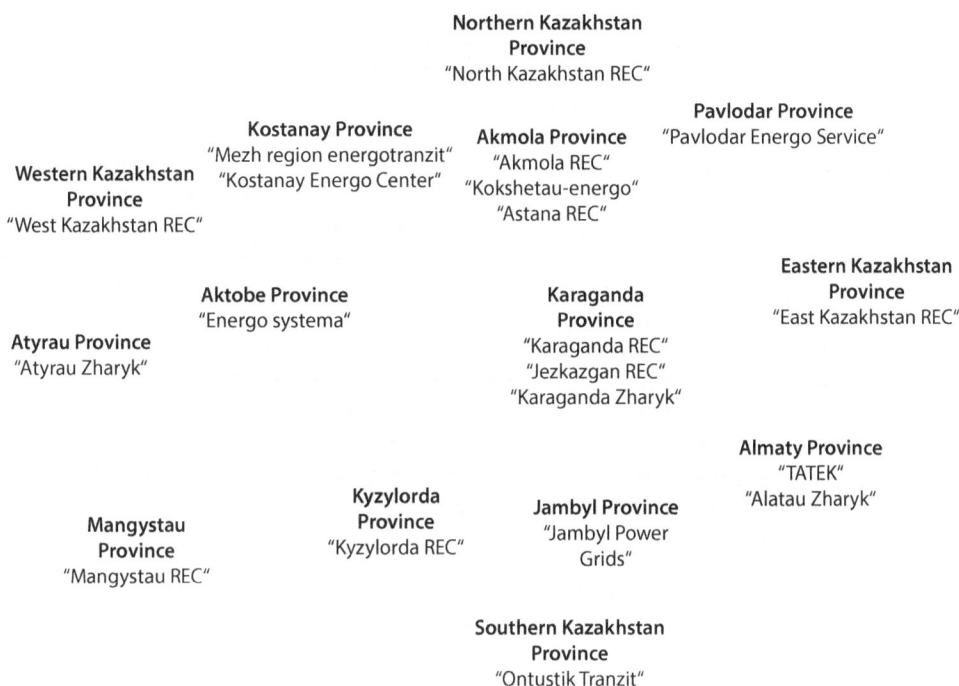

Northern Kazakhstan Province
"North Kazakhstan REC"

Pavlodar Province
"Pavlodar Energo Service"

Kostanay Province
"Mezh region energotranzit"
"Kostanay Energo Center"

Akmola Province
"Akmola REC"
"Kokshetau-energo"
"Astana REC"

Western Kazakhstan Province
"West Kazakhstan REC"

Eastern Kazakhstan Province
"East Kazakhstan REC"

Aktobe Province
"Energo systema"

Karaganda Province
"Karaganda REC"
"Jezkazgan REC"
"Karaganda Zharyk"

Atyrau Province
"Atyrau Zharyk"

Almaty Province
"TATEK"
"Alatau Zharyk"

Kyzylorda Province
"Kyzylorda REC"

Jambyl Province
"Jambyl Power Grids"

Mangystau Province
"Mangystau REC"

Southern Kazakhstan Province
"Ontustik Tranzit"

systematic database records the quality and reliability of electricity supply at the retail level. Distribution companies are not required to calculate internationally accepted standard measures to assess the reliability of supply: the System Average Interruption Duration Index and the System Average Interruption Frequency Index. However, anecdotal evidence suggests that there are significant problems with the quality and reliability of power supply, especially in rural regions.

In 2004, in an attempt to break up the vertically integrated regional monopoly of the RECs and kick-start retail competition, the government undertook a full unbundling of distribution by splitting the supply function from the existing RECs and allowing the operation of stand-alone electricity supply organizations (ESOs), which account for the bulk of electricity sales at the retail level. The unbundling is a full legal and functional separation (that is, the network, generation, and supply companies are legally stand-alone entities). However, the companies may belong to one owner (for example, a REC) within a holding structure. Network companies cannot engage in the purchase and sale of electricity, except for buying electricity to cover their network losses. By law, RECs are required to provide nondiscriminatory access to their networks and are considered natural monopolies and regulated.

Stuck in Transition • http://dx.doi.org/10.1596/978-1-4648-0971-2

There are about 180 ESOs—many of them privately owned—of which about 40 are large enough to enjoy natural monopoly status subject to regulation. Some supply power to residential users while others focus on servicing commercial and industrial customers. In the retail sector, customers consuming 1 MW of electricity or more have the right to buy electricity directly in the wholesale market—from generators or the spot market—or from an ESO. Many large customers have exercised their choice and, hence, the large-customer supply segment is relatively competitive. However, smaller customers—although also legally free to change their supplier—remain, in practice, tied to the local ESO. In many regions, the local ESO controls nearly 100 percent of the respective market. In this segment of the retail market, customer choice exists largely on paper.

The lack of automated commercial metering and communications systems makes retail access technologically very difficult. This circumstance also handicaps regional ESOs from entering the national wholesale market and other regional markets through KEGOC's transmission system. An additional factor limiting retail choice is the effective lack of transparent and enforceable grid access rules at the distribution level, notwithstanding legally allowed open access.

Because of its excessive fragmentation, often nontransparent ownership, politically influenced tariff setting, grossly insufficient funds for modernization, and inefficient operation, distribution has become the weakest link of the Kazakhstani power sector. This is reflected by its high share (57 percent) of outdated equipment and excessively large network losses, amounting to 13 percent on average (KazEnergy 2013). In the past two decades, the subsector has been in a cash-strapped and virtual survival mode.

Although quality-of-service complaints are widespread, the fragmented nature of the distribution sector only compounds the problems. Final consumers have contractual relations only with ESOs, which cannot own and operate low-voltage equipment (such as distribution lines, transformers, and meters), which belong to the RECs. However, the RECs do not interface directly with consumers and do not have any contractual obligation to address their complaints. Because of this divided responsibility among RECs and ESOs, final consumers often remain unprotected and poorly served. Another shortcoming of the retail market is that several ESOs are closely associated with generators as their marketing subsidiaries, and thus lack a strong interest in shopping for the cheapest power in the national market, thereby limiting competition.

Smart grids and smart meters at the distribution level are now under consideration for adoption.[20] With the government's new emphasis on the Green Economy Concept, including more intensive utilization of the country's abundant renewable energy resources, a directional shift toward decentralized generation looks unstoppable. Smart grids are particularly suitable to enable local, small-scale renewable energy–based power production and its integration on the grid in a bidirectional manner.

District Heating

A sizable chunk of Kazakhstan's electricity production is generated at CHPs, with a total installed electricity capacity of 7 GW (map 3.5). Some CHPs are privately owned, but most are under state and municipal ownership. Forty percent of the country's thermal energy for heating comes from CHPs and through the centralized heating networks in big cities. Most of the existing CHP capacity in the country was built between 1960 and 1980. As a result, operating performance characteristics (reliability and heat rates) are poor. For example, the available electricity generating capacity is only about 83 percent of the design capacity. At 23.4 percent, the gap is even higher on the heat side (KazEnergy 2014, 164). Because of the large-scale technical obsolescence, the CHP sector faces a massive need for modernization and retirement. Under the government's energy strategy, CHPs are expected to remain an important part of the country's power supply. Currently, around 80 percent of CHP capacity runs on coal. This is a sensitive environmental issue, because CHPs are located near (or in) cities and have insufficient emissions control equipment. For that reason, a shift from coal to natural gas for CHP generation in the future might become a strategic target.

In 2013, total generated heat amounted to 111.7 million gigacalories (GCal), of which 30 percent was produced by industrial plants. Of the remaining

Map 3.5 Distribution of District Heating Generation across Kazakhstan, by Boiler Houses and Combined Heat and Power Plants

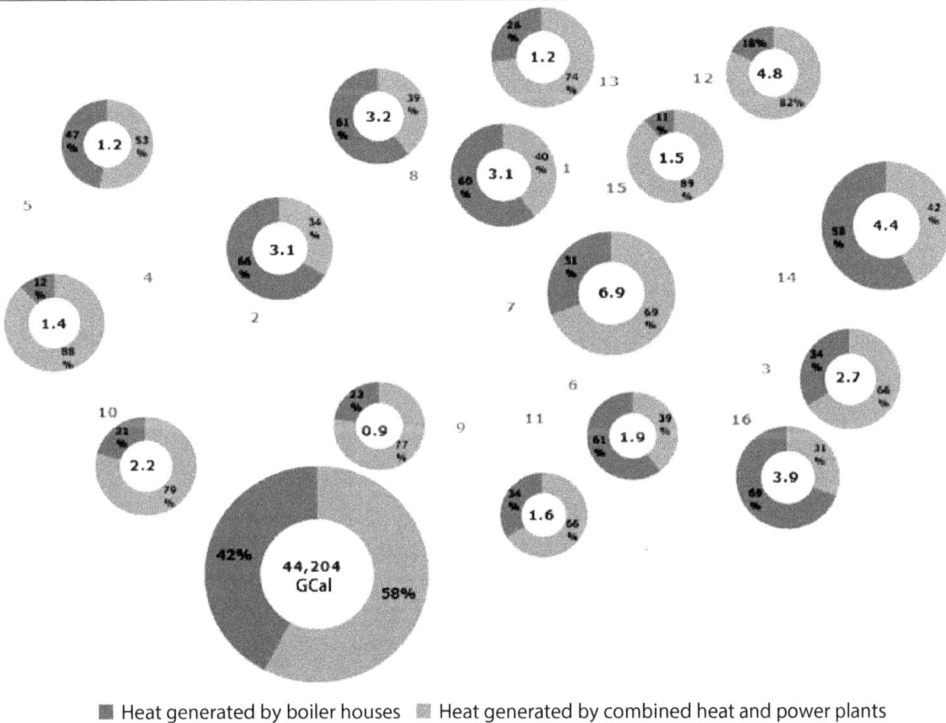

■ Heat generated by boiler houses   ■ Heat generated by combined heat and power plants

Note: GCal = gigacalories.

77.8 million GCal, approximately 50 percent was produced by CHPs, 20 percent by other coal-based TPPs, and 30 percent by boiler houses. Of the 111.7 million GCal of generated heat, 12.4 million GCal was used for electricity production and 16.0 million GCal for own needs. Significant cross-subsidization occurs between heat and electricity tariffs, with shares of fuel and operating costs allocated disproportionately to the latter. KazEnergy predicts a 1.7 percent increase in the thermal requirement in Kazakhstan per year until 2030 (figure 3.6).

The regulatory framework for CHPs does not provide a consistent and supportive environment for their optimal development. Although the price for byproduct heat—a natural monopoly service—is regulated, the price of electricity is determined in a competitive wholesale electricity market. In a cogeneration plant, it is inherently difficult to distinguish between power costs and heat costs. Under the prevailing regulatory procedure, the regulator, for social reasons, tends to allocate the bulk of the total costs to power generation. Generally, the heat business is loss-making because of regulated end-user tariffs, which are kept low. Furthermore, if the power load is not sufficient, as is typical in the nonwinter seasons, CHPs have to switch to condensing mode, which has very low efficiency compared with the conventional condensing power plants. The resulting cross-subsidization of heat generation by power generation substantially overstates the actual cost of electricity, thereby rendering CHPs less competitive in the wholesale electricity market. As a result of these factors, the CHP sector faces difficulties in attracting investors for new plants. However, the largest CHPs were included in the government's electricity sector program, "Tariffs for Investments," which allowed annually escalating price caps for CHPs in 2009–15, including a component for return on investment, which was agreed with the government.

Figure 3.6 Heat Energy Production, by 2030

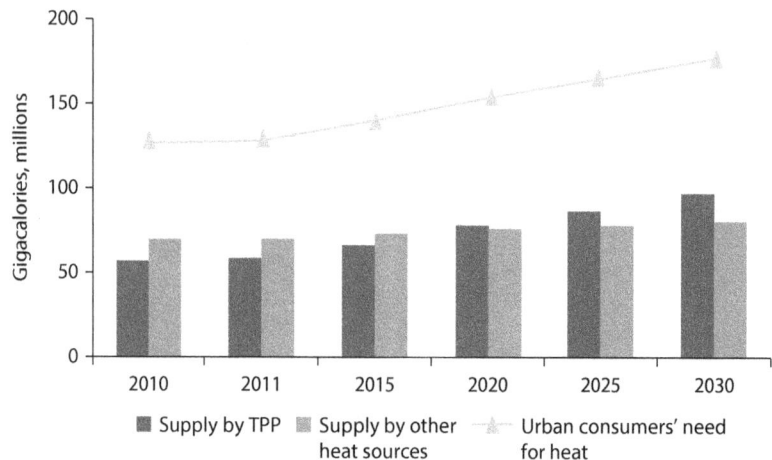

Source: KazEnergy 2013.
Note: TPP = thermal power plant.

Renewable Energy Framework

Organizational and Contractual Structure

The organizational and contractual arrangement of Kazakhstan's renewable energy electricity sales differs from that of its wholesale market. While the latter's structure is based on bilateral contracts between generators and large consumers, as well as ESOs, laws relating to renewable energy require the sale and purchase of renewable energy electricity by way of a single buyer, the Financial Settlement Center (FSC). The FSC must buy not only renewable energy electricity, but all power from renewable energy operators under 15-year power purchase arrangements. The FSC then blends the renewable energy electricity to onsell to conventional generators (conditional consumers). Should any of these consumers cancel for whatever reason, however, renewable energy payments to the FSC must be redistributed among the remaining conditional consumers for the entire amount of the renewable energy needed. To achieve a fair and equal distribution of costs for all customers, the conditional consumers would onsell the renewable energy electricity to large consumers and ESOs, together with their own energy. Given the tariff cap regulation for conventional generators, the transfer cost to the final consumer must be assured by taking into account additional costs, although this is currently not explicitly required by the regulators.

The sale and purchase of renewable energy electricity through a single buyer is not an unusual practice; it takes place regularly to form market pools. In Kazakhstan, however, two factors make the contractual structure atypical and, to a certain extent, somewhat a challenge:

- The absence of a pool market, whereby electricity is usually sold according to direct bilateral contracts. The FSC is therefore a structure that will purchase renewable energy and, unlike other renewable energy single buyers—which are often state-owned enterprises (for example, Ukraine) or parts of the system operator (for example, the former Yugoslav Republic of Macedonia and Serbia)—it is a weak entity without a history and is not adequately integrated into the market framework. The FSC does not have the required short- and long-term creditworthiness necessary for long-term PPAs with international banks.
- The involvement of conditional consumers (that is, conventional generators) in the payment chain and planning, scheduling, and balancing phases of renewable energy is unprecedented at the global level. In the case of Kazakhstan, it was considered easier to collect the renewable energy support tariff from conventional generators than directly from large consumers and ESOs. Although there may be some merit to this, the design of this contract model reflects the lack of understanding of electricity market structures, and it has led to issues in the interaction between the renewable energy and wholesale markets.

The organizational and contractual structure of renewable energy sales in the Kazakhstani market poses many challenges. Consideration should be given to implementing additional measures for market integration, reviewing the legal setup of the FSC, or providing additional guarantees.

Feed-in Tariffs

Feed-in tariffs—referred to in Kazakhstan as fixed tariffs and opted by Kazakhstan to promote investment—originate from a benchmarking approach, where fixed tariffs are set based on feed-in tariff levels in select countries with similar conditions. Although the nontransparent nature of the commonly used fixed tariffs has been questioned by some investors and experts, they have been generally accepted by international financial institutions.

The lifespan of a fixed tariff is 15 years, in line with international practice. Separate fixed tariffs apply to different technologies (for example, wind, solar photovoltaic, hydro, and biogas), and are adjusted annually for inflation once an installation is commissioned. The fixed tariff scheme does not differentiate between project size or resource availability; that is, it takes into account neither economies of scale between projects that vary in size nor projects of a lower unit cost in more attractive locations. At the inception stage of a feed-in tariff scheme, however, it is not unusual to have a slight degree of differentiation.

Kazakhstan's fixed tariff scheme currently has no restriction on eligibility. Renewable energy projects that fulfill basic registration conditions are added to the project list for the fixed tariff scheme. Concerns by the government about too rapid and high an inflow of renewable energy projects may result in some reviews and/or limitations, although this has not yet been confirmed. However, the government has admitted that for some projects, fixed tariffs have not been adequate, and it has arbitrarily approved separate fixed tariffs for locally produced solar photovoltaic and wind plants relating to EXPO 2017.[21]

Although these special fixed tariffs are moderate in volume (37 MW for locally produced solar photovoltaic and 100 MW for wind), they are nevertheless substantially higher than the usual fixed tariffs for these technologies by approximately three times. Such high fixed tariffs are of concern because of their ad hoc and, hence, discriminatory nature. The legality of these special fixed tariffs also raises doubts.

Renewable Electricity Network Connection

In principle, the current legal provisions provide for guaranteed network connection, stipulate that network companies must connect any renewable energy electricity producer anywhere, and establish that the need for network upgrading is not a valid reason for refusal. Furthermore, renewable energy producers should only pay for connecting their installation to the "closest connection point" on the network, without liability for other costs (that is, "shallow pricing" in the literature). In practice, however, significant issues arise in connecting Kazakhstani renewable energy producers:

- Despite the fact that the shallow price principle costs that are necessary for upgrading should be recovered from the network tariff and renewable energy producers, in practice, this cannot be effected by the network company because of the current regulatory requirements.

- At present, there is a need for a formal connection agreement, and the PPA does not include the right to connect.
- There are no specific technical rules for renewable energy in the Grid Code, thus permitting network companies to issue impossible requirements.

These renewable energy connection issues are not trivial, and constitute a real barrier to the development of renewable energy in Kazakhstan. Although the legal rights of renewable energy operators to be connected are unprecedentedly favorable, in practice, network companies are in no position to fulfill their legal obligation.

Scheduling, Dispatch, and Balancing

Renewable energy producers should be required to fulfill the same scheduling timelines as do other producers. The rules of centralized purchase and sale by the FSC require renewable energy generators to provide monthly, daily, and hourly power supply schedules for the following month, far in advance of actual production. Although renewable energy producers are not liable for forecast inaccuracy, they do, nevertheless, provide daily forecasts.

Since the FSC has a purchase and offtake obligation for the power produced, renewable energy producers should have prioritized dispatch, according to the legislation. However, the dispatch is more an implicit notion than an explicit obligation for the system operator and network operators. This notion is reinforced by the absence, in the contractual structure, of suitable relationships between the renewable energy operator and the system operator and network operators. In the current legal environment, generators generally have so-called dispatch contracts rather than connection contracts, the former of which are unsuitable for renewable energy and do not relate to prioritized dispatch or guaranteed offtake. Therefore, it is unclear how the prioritized dispatch would operate on the network under normal circumstances or irregular conditions (for example, network congestion). There are no clear procedures in place should renewable energy be downturned and redispatched, or for the extent to which renewable energy would be compensated.

The exemption of renewable energy operators from direct balancing costs is a concrete part of the Kazakhstani renewable energy support framework. The additional balancing costs incurred by the system or other market participants should therefore be covered.

Tariff Regulation and Subsidies

Tariff regulation was formerly carried out by the separate AREM. In 2015, AREM was transformed into the Committee for Natural Monopolies and Protection of Competition under the Ministry of National Economy.

The state-controlled regulatory system has evolved over the past two decades. However, sector regulation still lacks an adequate degree of autonomy and is highly vulnerable to political interference at the national and regional levels, as well as to varying degrees of "regulatory capture" by a powerful incumbent sector and political entities.[22] In tariff regulation, the Energy Concept 2030 places a disproportionate emphasis on moderating tariff increases instead of presenting a time-bound transition path to a sound, professional regulatory framework, including performance-based tariff regulation.

A professionally managed and reasonable autonomous regulatory institution is crucial to (a) improve the performance of the natural monopoly segments of the power sector; (b) support competition in the contestable market segments; and (c) attract investors to the sector, including the high-potential renewable energy sector. International good practice shows that a well-designed and credible regulatory system reduces the cost of private capital for the power sector. This is an important benefit to reap in the capacity-constrained context of Kazakhstan.

Electricity tariffs best serve the public interest when they are established through a transparent, accountable, and participatory process. Procedural clarity involves identifying legal frameworks; key decision makers and procedures for setting and revising tariffs; and procedures and forums that allow consumers and other stakeholders to participate in decisions, appeal decisions, and seek redress of grievances. Kazakhstan is still a long way from such a tariff framework.

Historically, the tariff system was a traditional "cost-plus" type. A well-known shortcoming of this method is that it tends to dull incentives for regulated firms to minimize costs and improve service quality. This deficiency is observable in Kazakhstan across the entire electricity value chain. Although in some areas regulation has performed reasonably well (for example, transmission tariff setting), elsewhere it has remained weak (for example, retail tariff regulation and monitoring/enforcing nondiscriminatory third-party access to the grids), leaving large scope for regulatory discretion and political interference. Some important regulatory aspects—such as quality-of-service standards, market power monitoring, cross-subsidization between regulated and competitive services, and dispute resolution—continue to receive inadequate attention, including in the Energy Concept 2030. With its limited and technically underskilled staff, AREM has encountered major regulatory challenges.[23]

Generation Tariff Regulation: Stepping Back into the Past

Generation tariffs have been regulated administratively since 2009 under tariff caps (that is, maximum or "investment" tariffs). Under a complicated and not fully transparent scheme, generators were placed into 13 tariff groups on the basis of plant type, fuel used, and distance from fuel source, with an escalating upper limit (cap) set for a seven-year period (2009–15), to be adjusted annually.

All generators were legally mandated to develop a medium-term investment program. Each generator has undertaken a specific investment program in return for a negotiated tariff increase not to exceed the cap for the given tariff group. As expected, actual tariffs moved relatively close to the cap. In 2015, this system was extended for another seven years and three more tariff groups were added (appendix A).

Transmission Tariff Regulation

The transmission tariff is approved by the regulator on a justifiable cost-plus basis, including costs of government-approved new investments. Conceptually, the tariff approximates long-run marginal costs. The tariff is partially and transparently unbundled into wheeling, dispatch, and selected (but not all) ancillary system services.

Under the World Bank–funded North–South Electricity Transmission Project, Kazakhstan adopted modern zonal tariffs to approximate the geographically varying financial value of the transmission services provided. The zonal tariffs were to provide correct signals for the siting of new generation and transmission, and to foster competition among generators by reflecting existing deficits and surpluses among the country's power regions. Such tariffs are particularly suitable for large and geographically fragmented power systems such as Kazakhstan's. Instead of fine-tuning a conceptually correct pricing system, however, AREM abolished it (following the closure of the World Bank–funded project), claiming that network expansions (including construction of the second North–South interconnector) made zonal tariffs superfluous. A nationally uniform tariff (embodying the inefficient "postage stamp" concept) was reintroduced.

Retail Tariff Regulation

Retail tariffs are complex and vary significantly by region, consumer category, and time and volume of consumption. The old system (prior to 2013) was over-regulated, lacked a sound market rationale, and was not always applied consistently across all the regional branches of AREM, which were particularly vulnerable to local political pressures and "regulatory capture." In addition to households, tariffs are regulated for all distribution-level consumers who purchase power from ESOs. Simple two-part time-of-use pricing (daytime and nighttime tariffs) is applied to households, and three-part pricing (adding peak tariffs) is applied to legal entities. Regulated retail tariffs are regionally differentiated, reflecting the significant underlying cost differentials (figure 3.7).

In January 2013, tariff regulation for RECs shifted from the cost-plus approach to a benchmarking method to strengthen their generally poor operational and financial performance. The financial and operational indicators of peer companies were used for the performance benchmarking of RECs. Each REC was then assigned a task to improve its performance by including investment project costs in the company's tariffs, which were approved for a period of three years with the possibility of annual adjustments. For a variety of reasons—including the

Figure 3.7 Residential Electricity Tariffs, by Region
U.S. cents/kilowatt hour, without value-added tax

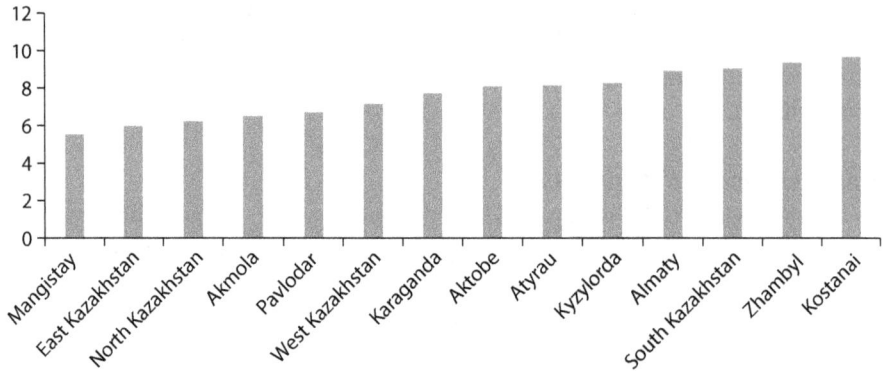

Source: Ministry of Energy.

difficulty of finding the right subsector "leader" for benchmarking, arbitrary regulatory decisions about the efficiency factor ("X factor"), lack of a meaningful investment component in the final tariff, and the poor overall financial performance of most of the RECs—in August 2015, the government decided to abolish the benchmarking system as unsuccessful. Therefore, as of January 2016, distribution tariffs switched from the benchmarking methodology to the controversial tariff cap-plus-investment commitment system that has been applied in the generation sector since 2009.

For ESOs classified as natural monopolies, the regulator applies a supply surcharge. Primarily for social reasons, however, the regulator frequently and arbitrarily freezes the retail tariffs, even when prices increase in the wholesale market.[24] This action hampers competition in the retail sector and threatens ESOs with bankruptcy.[25]

At about 2.3 U.S. cents/kWh, the distribution margin is compressed, which accounts for much of the poor financial performance of the subsector.[26] The recent large devaluation of the tenge further weakened the credit profiles of several RECs, due to a currency mismatch between their debt and revenue and the absence of hedging to reduce foreign exchange exposure. The evolution of the average residential electricity tariff is reflected in figure 3.8.

Looking ahead, the major regulatory challenges to address are to:

• Review the overall institutional framework, with the primary objective of developing a streamlined and efficient regulatory system, which would attract much-needed new investment and international expertise to the power sector. Regulatory capacity should be considerably strengthened and the regulator's autonomy increased to attain credibility among market participants. The regulator's government structure (terms of employment of key staff, structure of the board, and budget) should be updated in line with best practice.
• Introduce across the electricity value chain incentive (performance-based) regulation in place of cost-of-service regulation.

Figure 3.8 Average Residential Electricity Tariff, 2007–14

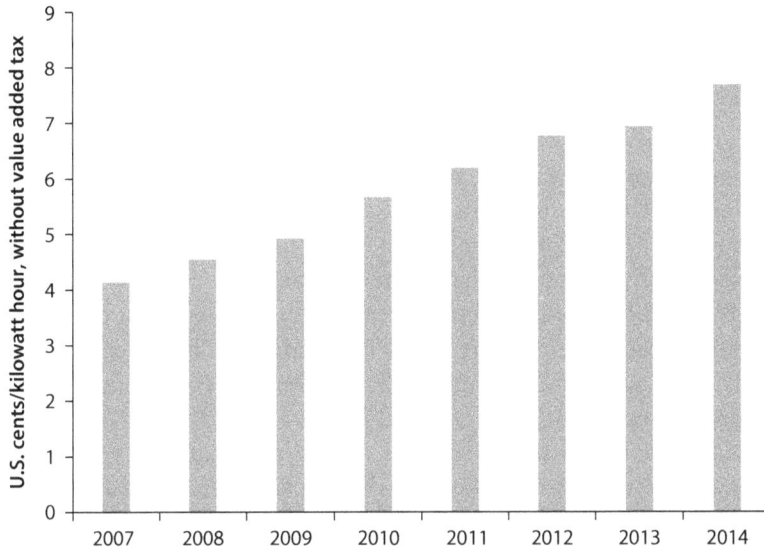

Source: Ministry of Energy.

- Adopt prospectively locational (nodal or zonal) transmission price signals to reflect accurately the cost of delivered energy and avoid "socializing" the cost of compensating for grid constraints (as is currently done under the nationally uniform transmission tariff). Nodal pricing would set higher power prices for delivery to congested load areas.
- Rebalance the tariff structure by more closely aligning regulated retail tariffs with the cost of wholesale power. Regulation should avoid retail rate freezes that expose distributors to an unsustainable squeeze on their cash flow when rising wholesale costs approach (or possibly even exceed) fixed retail rates.
- Discontinue regulatory setting of retail tariffs, starting with nonresidential customers, once a mature, competitive retail market develops.
- Introduce dynamic (fully differentiated time-of-use) pricing at the retail level, first for large consumers, followed by residential consumers, subject to mandatory installation of hourly digital meters ("smart meters").
- Establish vigilant market monitoring and effective controls on market power across the sector value chain.
- Design appropriate social protection mechanisms (for example, lifeline tariffs for the poorest consumers), based on best practice, to manage the social consequences of tariff rebalancing.
- Lay out a detailed implementation road map for the implementation of renewable energy electricity, addressing the assignment of real financial resources to renewable energy electricity support costs, as well as the final reflection of these costs—including transmission investments—in retail tariffs. Address the uncertainty over the creditworthiness of the FSC and the bankability of the PPAs between developers of renewables and the FSC.

Fossil Fuel Subsidies

In 2013, Kazakhstan had an average energy subsidization rate of 32.8 percent; subsidies per capita were US$359; and the total subsidy was 2.8 percent of gross domestic product (GDP) (table 3.1). Fossil fuel subsidies fell, from US$9.1 billion in 2011 to US$6.1 billion in 2013 (table 3.2), and compare well with those in other oil-rich countries (the Islamic Republic of Iran, Libya, Mexico, Nigeria, and Russia) and some oil importers (Argentina, India, Ukraine, and Uzbekistan). In Kazakhstan, most of the subsidies were directed to oil (US$2.1 billion) and coal (US$2.4 billion), with electricity and gas receiving less than US$1 billion each. Indirect budgetary energy subsidies are significant and widely used, although direct budgetary support for electricity and heat consumers has been largely eliminated. Support is provided indirectly, however, by regulating electricity and heat tariffs. Price caps still exist for most fuels, causing significant market distortions.

Although the International Energy Agency estimated a relatively large overall implicit energy subsidy for Kazakhstan (2.8 percent of GDP), there is little evidence that the power sector has been a major recipient of this subsidy. Much of coal mining is privately owned, and coal is naturally cheap for the mine-mouth generators in the Ekibastuz power generation hub. Coal is a tradable commodity with a market-clearing price in Kazakhstan. The only major issue is the lack of cost internalization of the considerable negative environmental externalities caused by the coal industry and coal-fired power generation.

Table 3.1 Average per Capita and Total Fossil Fuel Subsidies

	Average subsidization rate (%)	Subsidy per capita ($/person)	Total subsidy as share of GDP (%)
Kazakhstan	32.8	359	2.8

Source: IEA 2013.
Note: The average subsidization rate is as a proportion of the full cost of supply. GDP = gross domestic product.

Table 3.2 Fossil Fuel Subsidies, by Fuel Type, 2011–13

	2011	2012	2013
Oil	2.5	1.7	2.1
Electricity	2.5	2.4	0.7
Gas	0.9	0.8	0.8
Coal	3.2	2.8	2.4
Total	9.1	7.7	6.1

Source: IEA 2013.

Notes

1. The thermal electricity breakdown is 84 percent coal, 14 percent gas, and 2 percent oil.

2. Approximately 90 percent of known coal resources are concentrated in the northern and central parts of the country.

3. Average ash content for Kazakhstan coal is 29 percent. Average heating value in the Ekibastuz basin is 3,900 kilocalories/kilogram; in the Karaganda basin, it is 5,200 kilocalories/kilogram.

4. Letter of approval for the General Scheme of Gasification of Kazakhstan for 2015–30, submitted by Prime Minister Karim Massimov and approved by Government Resolution of the Republic of Kazakhstan in 2014.

5. Only 20 percent of the total volume of gas transported in Kazakhstan is actually consumed in Kazakhstan; the rest is transferred to China and Russia through a network that includes Turkmenistan and Uzbekistan.

6. The scope of corporate governance includes the existence of a corporate governance code, internal bylaws governing operations of the company's key governance bodies (a general meeting of shareholders, the board of directors, and the executive body), and antifraud and anticorruption procedures.

7. These include the Aksu power plant owned by ENRC, Temirtau power plant owned by Arcelor Mittal, Aktobe power plant owned by Kazchrome, and Bukhtyrma hydropower plant owned by KazZink.

8. There is an ongoing debate in the international professional literature about whether "energy-only" liberalized markets could send sufficient signals in time to mobilize the necessary investment as capacity margins tighten. The differing opinions on this subject were at the root of the divergence of views as to the adequacy of the capacity margin. Experience across Organisation for Economic Co-operation and Development countries has shown that well-functioning markets, where signals are undistorted in effectively liberalized markets and companies have incentives to invest, can deliver appropriate and timely signals for new power generation investment. However, underinvestment, supply shortages, and price spike crises have been seen in some markets (for example, Ontario, Canada, in 2002–03, and Victoria, Australia, in 2000) where political and regulatory uncertainty obscured the price signals available to investors in the electricity market. In large-scale, engineered systems, even a temporary period of uncertainty can have more amplified effects than in other commodity markets, because of the long lead times and investment risks associated with planning and building new generation investments (IEA 2005; RAE 2013).

9. The generation capacity margin is defined as the excess of available capacity over the peak demand (as a percentage of available capacity).

10. For example, during winter 2008–09, power supply restrictions were in the range of 230–360 MW in southern Kazakhstan, a high power deficit in part of the country.

11. For example, at 32 percent, the average operational efficiency of the Kazakhstani coal-fired condensing power plants is very low, compared with 42 percent in the best-performing systems abroad.

12. The first 660 MW unit was supposed to come on line by 2015. However, because of many problems, as of mid-2016, construction of the power plant had not started in earnest.

13. See http://ranking.kz/infopovody/nergetika_kazahstana_perezhivaet_investicionnyj_podem, May 25, 2015.

14. Unofficially, government officials noted in mid-2016 that the extended tariff cap program will not involve investment commitments on the part of the generators, and restrictions on the use of profits will be lifted. If these indeed hold, then the tariff caps become merely a de facto anti-inflation device. Clearly, this is a step in the right direction.

15. For example, because of large-scale modernization and extension investments supported by the World Bank, the frequency of major transmission outages dropped from 29 in 2009 to four in 2013. The total duration of outages per year fell dramatically, from 111 to two hours during the same period.

16. In a high-profile, open-access case, KEGOC had to pay considerable damage compensation to a major foreign-owned generator, which complained about the lack of non-discriminatory access to reach export markets through the national grid. In the absence of an appropriate internal dispute resolution mechanism (including penalties levied on KEGOC in proven cases of discriminatory access management), the generator had to resort to international arbitration. Later, the firm divested itself from thermal generation in Kazakhstan.

17. KEGOC transmits less than half of the electricity (45 percent in 2013) generated in Kazakhstan. A significant portion of electricity is transmitted within vertically integrated regional systems.

18. Transmission constraints are understood to include line flows, bus voltages, equipment ratings such as transformer tap limits, and generation limits on active and reactive power, among others.

19. Especially for foreign-owned generators, this is a sensitive issue because in some national systems, particularly vertically integrated ones, it was observed that system operators may issue biased estimates of the real-time availability of transmission capacity, make questionable curtailments of transmission service for system security reasons, or reserve excessive allocations of transmission import capacity for themselves.

20. The term "smart grid" here means combining differentiated, time-based power prices with information technologies that can be set by users to control their use and/or self-generation automatically, lowering their power costs and offering other benefits, such as increased reliability to the system as a whole.

21. EXPO 2017 is an International Exposition scheduled to take place between June 10, 2017, and September 10, 2017 in Astana, Kazakhstan.

22. In Kazakhstan, social protection is tied to the prices of utility services. If electricity prices rise and the cost of utility services exceeds a defined percentage of the income of low-income families, local governments have to provide additional social cash transfers. Thus, price controls effectively move a part of the cost of social protection to utilities, which is unsustainable in the long run. Some limited social protection is also provided under the two-block residential tariff system for households, whereby tariffs for the basic level of consumption are set somewhat lower than for the block above the basic level.

23. "Market power" (or "antimonopoly" in Kazakhstani parlance) monitoring of the power sector is performed by another institution, the Agency for Protection of Competition (known as the Anti-Monopoly Agency) This agency has limited qualified staff to conduct thorough investigations of market abuse across the entire national economy. Moreover, recently the agency seems to have focused on observed "speculative" market actions instead of preventing the structural emergence of market power, potentially leading to abusive anticompetitive conduct.

24. Electricity subsidies provided indirectly by regulating electricity tariffs, and which are kept below the full cost of the provision of service, were estimated at US$2.5 billion in 2011, US$2.4 billion in 2012, and US$0.7 billion in 2013 (IEA 2014).

25. ESOs have no control over more than 90 percent of the cost of electricity purchased on the competitive wholesale market from generators.

26. Estimated for 2014 on the basis of the Astana service area, which is served by Ekibastuz (excluding value added tax [VAT]):
Generation cost (Ekibastuz): 4.3 U.S. cents/kWh
Transmission margin: 1.1 U.S. cents/kWh
Distribution margin (split between the Astana-based REC and ESO): 2.3 U.S. cents/kWh
Retail tariff in Astana: 7.7 U.S. cents/kWh.

References

DNV GL. "Regulatory and Institutional Improvement for Renewable Energy Investment in Kazakhstan." DNV GL, Kazakhstan.

GoK (Government of Kazakhstan). 2014. "Energy Concept 2030." Concept on Development of the Fuel and Energy Complex of Kazakhstan until 2030.

IEA (International Energy Agency). 2002. "Tackling Investment Challenges in Power Generation in IEA Countries." OECD, Paris, France.

IEA (International Energy Agency). 2005. "Lessons from Liberalized Electricity Markets." IEA, Paris, France.

IEA (International Energy Agency). 2013. "World Energy Outlook 2013." OECD, Paris, France.

_____. 2007. "Tackling Investment Challenges in Power Generation in IEA Countries." IEA, Paris, France.

_____. 2013. "World Energy Outlook 2013." OECD, Paris, France.

_____. 2014. "World Energy Outlook 2014." OECD, Paris, France.

KazEnergy. 2013. "KazEnergy National Energy Report 2013," Astana, Kazakhstan.

KazEnergy. 2013a. "KazEnergy National Energy Report." Presentation (undated). Astana, Kazakhstan.

_____. 2013b. "The National Energy Report." Astana: KazEnergy.

KEMA. 2010. "Kazakhstan Generation Roadmap: Component 2: Conceptual Solution. Analysis of Adopted Capacity Market Model and Description of its Functioning." Astana, Kazakhstan.

KEMA. 2013. "Kazakhstan Generation Roadmap: Component 2: Conceptual Solution. Analysis of Adopted Capacity Market Model and Description of its Functioning," March.

Kikeri, Sunito, and Aishetu Fatima Kolo. 2005. "Privatization: Trends and Recent Developments." World Bank Policy Research Working Paper No. 3765. World Bank, Washington, DC.

McKinsey & Company. 2002. "Perspectives on Corporate Finance and Strategy."

RAE (Royal Academy of Engineering). 2013. "GB Electricity Capacity Margin: A Report by the Royal Academy of Engineering for the Committee of Science and Technology." RAE, London.

Sarbassov, Yerbol, Aiymgul Kerimray, Diyar Tokmurzin, GianCarlo Tosato, and Rocco De Miglio. 2013. "Electricity and Heating System in Kazakhstan: Exploring Energy Efficiency Improvement Paths." *Energy Policy* 60 (C): 431–44.

CHAPTER 4

Sector Strategy and Reforms

Sector Strategy and Market Development

Following independence, Kazakhstan became one of the first former Soviet Union states to embrace a market-oriented strategy to reform the ailing, dysfunctional electricity sector. Kazakhstan quickly became a frontrunner in staged sector reforms. To a large extent, radical market reforms were prompted by the sector's deep financial and operational crisis in the wake of the collapse of the Soviet Union and its interconnected power system. Under a pro-market strategy, the old vertically integrated state monopoly was broken into separate electricity generation, transmission, and distribution units. The bulk of electricity generation and a significant part of its distribution were privatized and a fiercely competitive wholesale market was introduced, relying on the enormous generation capacity surplus resulting from the collapse of electricity demand. During the protracted period of large excess capacity, however, the government of Kazakhstan's sector strategy lacked adequate foresight in one key area: the quality of legal and regulatory processes. This area became a major obstacle to investment across the entire electricity chain, except in the state-owned, high-voltage transmission segment. In that segment, upgrade and extension investments were largely funded by state-guaranteed loans from international financial institutions, including the World Bank. The poor investment climate led to several high-profile departures from the sector by foreign strategic investors (for example, AES Corporation and Tractabel) while practically no major modernization and expansion investments took place in generation and distribution.

Projected reserve margins became dangerously tight by the mid-2000s. The government faced a crucial dilemma: either (a) sustain the liberalizing reform agenda and fully open up the entire sector, including distribution, to genuine structural, regulatory, and pricing reforms, thereby generating strong enough financial incentives to investors; or (b) shift to centralization, renationalization, heavy-handed regulation, and state control across the entire value chain, including generation, to ensure supply adequacy and security. The high-inflation environment of this period resolved this dilemma, as the government was particularly concerned about the adverse inflationary and export competitiveness effects of potential runaway electricity prices amid the perilously tightening supply and demand balance. In its overreaction to signals of an impending generation crunch, the government's strategy shifted from market liberalization to intrusive state intervention and micromanagement. Especially under the 2008 amendments to the Electricity Law, many essential reforms were rolled back (for example, privatization, sector unbundling, liberalized wholesale market, spot market, and zonal transmission pricing) or delayed (for example, balancing market and distribution tariffs) (figure 4.1). Having identified as most critical the lagging generation investments and a capacity deficit with potentially damaging implications for the national economy, the government made crucial amendments to the Electricity Law in 2008. These were a game changer. Since then, the government has single-mindedly focused—at the expense of rolling back sector reforms—on achieving a steep increase in rehabilitation and expansion investments in generation. The authorities justified these emergency-type measures as unavoidable and of medium-term significance, at most.

The government of Kazakhstan's Energy Concept 2030 and the more recently adopted programs and policy measures[1] promise more of the same; that is, prolongation of intrusive, state-led sector management. Since the adoption of the Energy Concept 2030, the onset of the macroeconomic crisis, in the wake of the collapse of world oil prices since mid-2014, and the consequent large devaluation of the national currency may actually prompt the government to tighten

Figure 4.1 Evolution and Devolution of Reforms, 1995–2016

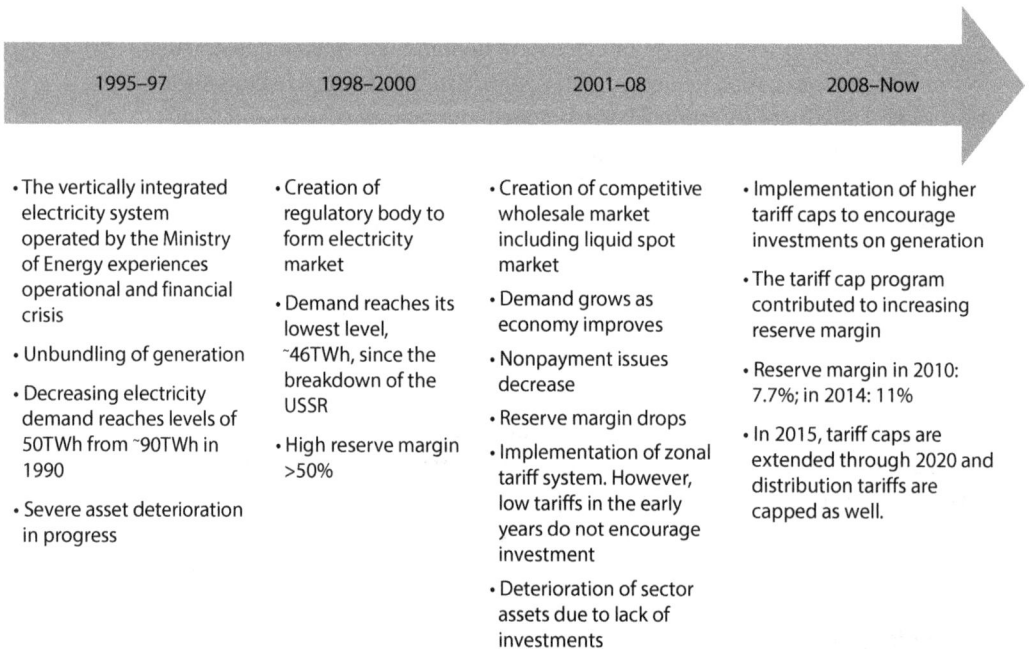

1995–97	1998–2000	2001–08	2008–Now
• The vertically integrated electricity system operated by the Ministry of Energy experiences operational and financial crisis • Unbundling of generation • Decreasing electricity demand reaches levels of 50TWh from ~90TWh in 1990 • Severe asset deterioration in progress	• Creation of regulatory body to form electricity market • Demand reaches its lowest level, ~46TWh, since the breakdown of the USSR • High reserve margin >50%	• Creation of competitive wholesale market including liquid spot market • Demand grows as economy improves • Nonpayment issues decrease • Reserve margin drops • Implementation of zonal tariff system. However, low tariffs in the early years do not encourage investment • Deterioration of sector assets due to lack of investments	• Implementation of higher tariff caps to encourage investments on generation • The tariff cap program contributed to increasing reserve margin • Reserve margin in 2010: 7.7%; in 2014: 11% • In 2015, tariff caps are extended through 2020 and distribution tariffs are capped as well.

Source: World Bank staff.
Note: TWh = terawatt hours.

further its administrative and regulatory grip on the sector, citing inflationary and social concerns.

The Energy Concept 2030 correctly identified the currently low level of competition in the generation sector—mostly because of the high share of state ownership and own generation by large industrial holdings—as a crucial obstacle to improved sector performance. Therefore, the Energy Concept 2030 calls for "systematic liberalization and development of competition." However, an implementation road map has yet to be fully fleshed out. Some recently announced measures and policies are in direct contrast to the principles of market liberalization and competition, such as postponement until 2019 of the launch of the Capacity Market, the post-2015 application of the controversial generation tariff caps, and extension of the same tariff concept to distribution.

Under the Energy Concept 2030, the government's overriding objective is to ensure energy independence and ease capacity tightness in generation. However, indications to date suggest that the government plans to achieve this objective essentially by way of command-and-control methods and continued governmental micro-meddling. As a result, the risk of departure from the government's declared energy strategy of promoting competition and the least-cost generation expansion path is high.

Figure 4.2 displays the structure of the existing electricity market model; the envisaged new model in the Energy Concept 2030 is shown in figure 4.3.

Figure 4.2 Existing Model of Kazakhstan's Electricity Market

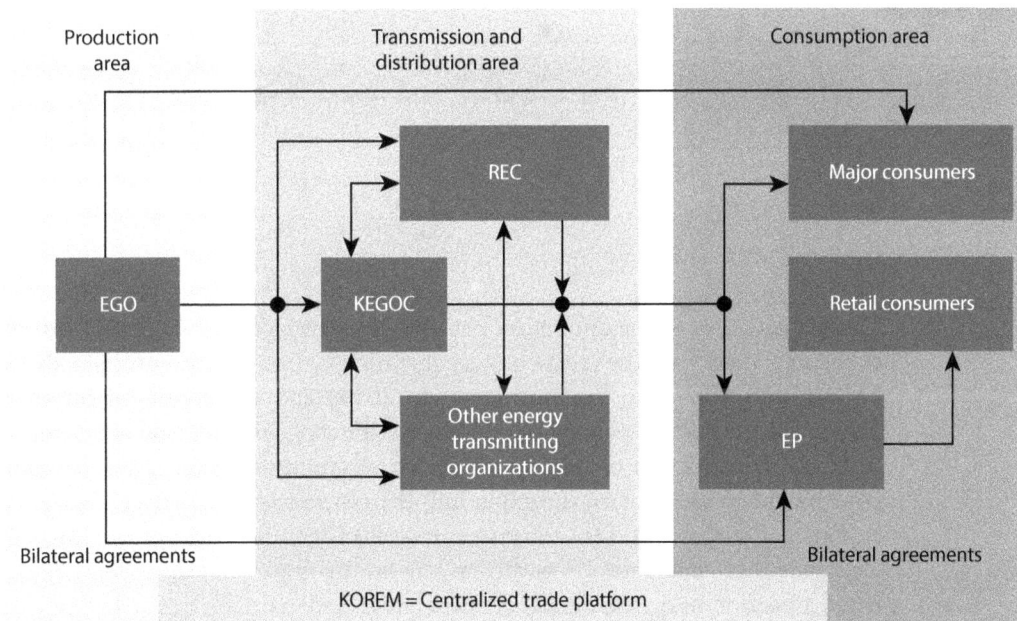

Source: Samruk-Energy 2014.
Note: EGO = energy-generating organization; EP = energy provider; KEGOC = Kazakhstan Electricity Grid Operating Company; KOREM = Kazakhstan Operator of Electric Power and Electric Energy; REC = regional electricity network company.

Figure 4.3 New Model of Kazakhstan's Electricity Market

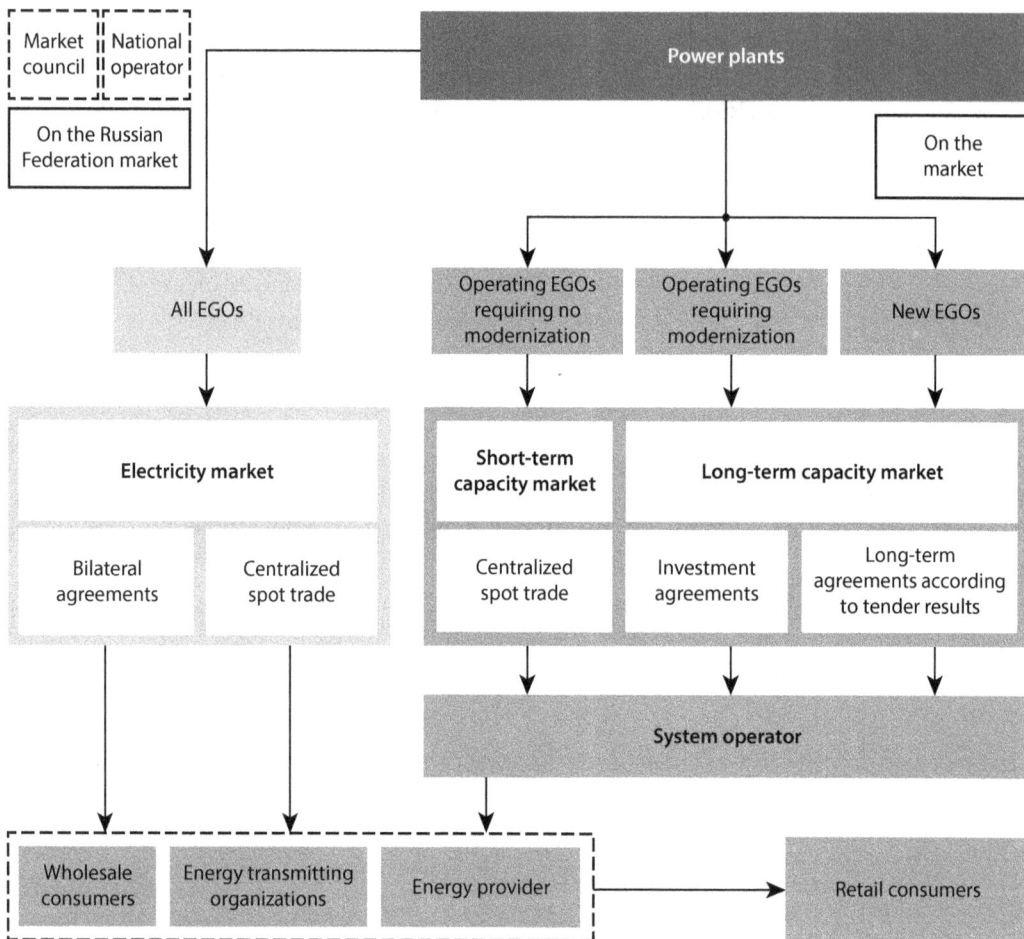

Source: Samruk-Energy 2014.
Note: EGO = energy-generating organization.

There are no conceptual differences between the two models. Two novel features of the new model are the creation of a generation Capacity Market (originally to be introduced in 2016 but postponed to 2019) and the establishment of a National Electricity Operator under Samruk-Energy with the following mandates: (a) construction of socially important electricity facilities if the relevant government tenders for construction fail, (b) centralized conduct of electricity exports and imports, and (c) construction of electricity facilities abroad. It is not clear why these mandates (in particular, the last two) are singled out for a designated operator in an essentially market-based and partially privatized power system. International trade and foreign investments in electricity should be treated as normal business transactions that are permitted to be conducted by all qualified participants in the power sector. The fact that the national operator is to be hosted within Samruk-Energy—the largest vertically integrated entity within the sector—further enhances the company's already sizable market power and is potentially detrimental to competition.

The government's energy strategy on renewable energy targets is extremely ambitious under the Green Economy Concept—30 and 50 percent of the generation in 2030 and 2050, respectively. However, these high shares are inconsistent with the government's desire to contain future tariff increases to maintain world market competitiveness for the electricity-intensive commodities that dominate exports. If the high renewable energy costs are not able to pass through fully to retail tariffs, how will the implied massive subsidies to renewable energy generators be funded? The Green Economy Concept is mute on this issue, other than to flag the call for "smoothing out" tariff increases through unspecified subsidies.

Electricity Market Structure: Rolling Back Reforms amid a Generation Capacity Gap

At present, the government's existing multimarket model—consisting of bilateral, spot, balancing, ancillary, and capacity submarkets—is grossly incomplete. Even the earlier well-functioning submarkets (bilateral contracts and spot) were damaged by the excessive state control and interventions imposed since the mid-2000s. The partial ancillary services market is managed by KEGOC, in a nontransparent manner. Introduction of the much-needed balancing market has been long delayed, and thus KEGOC "manually" balances the system in real time. The Capacity Market has already been designed, although it has considerable flaws that may compromise its effectiveness upon its introduction in 2019. Since the mid-2000s, from one of the most liberalized power markets of the former Soviet Union, Kazakhstan has steadily slid toward an oligopolistic generation structure that is dominated by a state-owned entity: Samruk-Energy. Renationalization and rebundling (reintegration) have proceeded hand in hand. The return of horizontal and vertical integration raises legitimate concerns about the possible abuse of market power. Intrusive and non-independent sector regulation has suffered a serious setback with the introduction of tariff regulation in generation by way of tariff caps. As a result, Kazakhstan's electricity market is now unattractive to foreign strategic investors. All major foreign investors have essentially exited from the sector. Competition in the wholesale market is more limited than previously, despite the fact that making better use of existing assets through competition is one way to delay the need for new generation capacity. Since effective competition puts pressure on companies to use capacity resources more effectively and practice just-in-time investment, it may allow the reserve margin to decrease without undermining system reliability.

Bilateral Contract Market

After Kazakhstan's independence, the wholesale market was fully liberalized based on bilateral sale/purchase transactions between generators and large consumers, as well as regional electricity companies and electricity supply organizations (ESOs), at prices agreed between the parties involved. Provided they acquired access to the transmission/distribution network, large consumers (> 1 megawatt [MW] daily usage) could enter the market and shop around for power. These direct consumers can contract with generators and KEGOC for their energy needs without engaging any distribution company. Bilateral contracts account for about 90 percent of traded volumes and, generally, function reasonably smoothly. The contracts—usually for a year—must contain an agreed hourly schedule. For such contracts, especially in the absence of effective balancing arrangements, the commitment to an hourly schedule is a considerable challenge for both sides. Until recently, generators usually offered to follow the load schedule of their customers and retailers. With the steadily narrowing capacity margin, however, some generators have started to be more restrictive in offering flexibility within the contracts. Recently, flexibility bands of only +/−5 percent have been reported by some generators. Bilateral contracts concluded by customers and retailers in the South and generators in the North are sometimes restricted by the transmission capacity of the North–South interconnector. KEGOC employs a nontransparent quota system to allocate congested capacity.

A major deficiency of a bilateral-contracts-only market is the lack of organized arrangements to manage imbalances that arise when actual generation and/or actual consumption differs from contracted volumes. In addition, bilateral contracts cannot ensure that network reliability is maintained and demand is served at least cost.

Until 2008, contract prices in the decentralized market were freely negotiated and confidential. The 2008 amendment to the Electricity Law, introducing generator price caps, fundamentally altered the bilateral contract market. The prices agreed in the contracts now may not exceed the regulated maximum tariffs set by the Ministry of Industry and Trade (now Ministry of Energy) not the sector regulator. In a further major restriction—and unlike other over-the-counter markets in the world—generator-to-generator and supplier-to-supplier trades are banned.

Spot Market

To address the deficiency of bilateral contracts, an electronic trading floor, operated by the Kazakhstan Operator of Electric Power and Electric Energy (KOREM), was launched in 2004—with considerable support from the World Bank under the Electricity Transmission and Rehabilitation Project with KEGOC—for short-term transactions, including intra-day, day-ahead, and other short- and medium-term deals. Participation in the market steadily increased until 2008, with more than 100 registered market participants and about 15 percent of all electricity traded in the KOREM market, providing valuable price signals about the wholesale market, including the rapidly tightening supply and demand balance. KOREM had become the most liquid, organized, short-term power market in the former Soviet Union (see figure 4.4).

Figure 4.4 Domestic Energy Trade, 2006–14

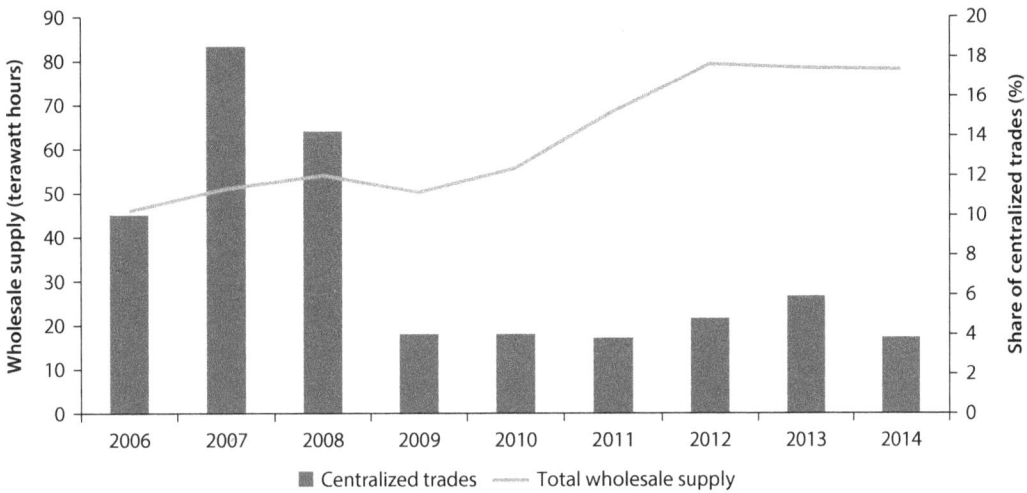

Source: KOREM.
Note: Centralized trades are spot market transactions under KOREM.

In 2009, KOREM's growth was undermined by a series of government decisions that included a ban on trader participation and inter-ESO transactions in the market, as well as the capping (at 10 percent) of the amount of electricity allowed to be sold on the spot market by generators under the tariff cap system—all but destroying the spot market. In recent years, KOREM has recovered somewhat, and in 2014 it posted a 7.4 percent market share. International experience shows that a well-managed and liquid spot market is an essential element of the overall wholesale marketplace, including for fine-tuning purchase and sales schedules, attracting private investment into generation without requiring special government guarantees, and integrating variable renewable energy into the wholesale market (for example, as in Germany).

Balancing Market

A real-time balancing market to manage supply and demand mismatches that naturally result from the bilateral contracts was fully designed and has been operated by KEGOC in trial mode since 2008. The government cited the following reasons for the long delay of live operation: (a) shortage of flexible (load-following) generation capacity to be called upon; (b) lack of sophisticated automated metering devices at generators to offer balancing services; and (c) perceived risk of insufficient load-following capacity, triggering excessive price hikes for balanced energy in a fully bid-based market. Although Kazakhstan is short of peak power, the existing system is unable to establish a representative price differential between base power and peak power to reveal the high real market value of the latter.

In the absence of a bid-based balancing market, KEGOC performs the balancing function manually, which is inefficient and nontransparent. The main balancing device is the utilization of interconnections, mostly with the Russian Federation. Internally, any market participant that has deviated from its schedule during one day (or hour) may be requested by KEGOC to deviate deliberately from the next day's schedule in the opposite direction. If necessary, KEGOC may enforce such deviations, for example, by limiting the consumption of a customer who has overconsumed the previous day, to "feed back" the "borrowed" power into the Russian system, or to compensate another customer who was regulated downward on the previous day. Clearly, this system is inefficient. In the absence of real-time operation on the balancing market, KEGOC's "manual" balancing does not generate valuable price signals to market participants to stimulate them to be in balance. The concept of "borrowing" power one day (hour) and "feeding it back" later is cumbersome, nontransparent, and somewhat arbitrary.

Ancillary Services Market

Under the Energy Concept 2030, the market model envisages the purchase of ancillary system services by the system operator from market participants. No organized market exists yet for ancillary services. Although this market is contestable in nature, KEGOC manages it essentially without regulatory oversight. Thus, the ancillary services market is a market only in name, since KEGOC contracts just a few operating reserves. Going forward, the government envisages two types of ancillary services to be purchased on a contractual basis by KEGOC: operating reserves and frequency control. Other ancillary services, such as regulation of reactive power for voltage control or black start capability, are not envisaged to be an explicit part of the future ancillary services market. Nevertheless, they should be provided on an obligatory basis, according to procedures set forth in the Grid Code. In particular, more attention should be paid to a market-based reactive power payment scheme to reduce substantially the currently large electric power losses on the grids. Increased reactive power consumption is typical of many industrial enterprises. Therefore, reactive power compensation is an important issue, because the reactive component is responsible for a part of the grid loss. At present, consumers pay for active power only, although it is reactive power that leads to partial grid losses. An effective payment scheme should therefore encourage consumers to find ways to compensate for reactive power, which will result in lower total power losses, voltage stabilization, and higher transmission capacity of networks.

Capacity Market

Revenue adequacy or "missing money" has emerged as a major problem in many organized wholesale electricity markets in the world. The "missing money" problem arises when the expected net revenues from sales of energy and ancillary services at market prices provide inadequate incentives for investors to generate new capacity. Many reasons explain why "energy only" markets often do not provide adequate incentives to invest in sufficient capacity. These include price caps that suppress prices below market-clearing levels, even during scarcity conditions, and out-of-market actions by government authorities and the system operator. In Kazakhstan, these problems have been compounded by the large generation capacity bubble that was inherited from the former Soviet Union, which led to "throat-cutting" competition among generators for cash-paying customers. For a long time, this allowed for covering only the short-run marginal costs of generation. Considering Kazakhstan's massive need for modernized and incremental generation, the introduction of an organized Capacity Market is an appropriate institutional response to stimulate large-scale expansion of generation capacity on a least-cost basis. A major advantage of the Capacity Market for investors is that it enables coverage of the capital costs of new capacity through selling (reserving) it to buyers (ESOs and large consumers) and receiving capacity payments, in addition to covering the variable costs through energy sales. When fully implemented, the Capacity Market should ensure an economically sound return on investment and provide incentives for construction of new generation assets or expansion of current capacity. A successfully functioning Capacity Market is likely to support the credit profiles of power generators.

The government intends to launch the Capacity Market in 2019, but important details (such as market rules and settlement arrangements) have yet to be fully finalized. Under the latest known version of the market concept, KEGOC would be the Capacity Market operator as a single buyer and seller of capacities, with ESOs and large consumers legally mandated to buy capacities covering their projected needs.

However, the market design has several flaws. It is unnecessarily complicated by envisaging two submarkets: a short-term (year-ahead) market for existing capacity and a long-term market for new capacity. By definition, a Capacity Market should focus on the long run; therefore, the need for the short-term submarket is open to question. Moreover, the existing design envisages continued use of the controversial generation price caps, thereby leading to overly administered and cumbersome arrangements that may undermine the integrity of the Capacity Market as a bid-based mechanism devoid of government micro-meddling.[2] Further, the potentially excessive financial liabilities of the market operator (KEGOC) under the Capacity Purchase Agreements raises legitimate concerns.[3] Therefore, additional provisions and measures are needed to safeguard the market operator's financial viability.

Single-Buyer Model?

Although the Energy Concept 2030 does not explicitly advocate the introduction of a single-buyer model (SBM), recent government pronouncements suggest a growing interest in its adoption.[4] Based on extensive international experience with SBMs, this would be an inappropriate policy choice for Kazakhstan.

As commonly understood, under the SBM a designated agency buys electricity from competing generators, usually has a substantial monopoly of transmission, and sells electricity to distributors and large power users without competition from other suppliers.[5] Although, internationally, the SBM has evolved under a variety of organizational forms and undoubtedly has some advantages,[6] the general experience, on balance, has been largely negative for several reasons. The SBM invites corruption, weakens payment discipline, potentially imposes large contingent liabilities on the government (mostly under the Power Purchase Agreements), creates financial losses for the single buyer when selling electricity at a below-cost regulated tariff,[7] responds poorly when demand falls—as in economic recessions—and hampers the development of cross-border electricity trade, thus leaving the latter to the single buyer (typically, a state-owned entity) without a strong profit motive.

Several countries that have introduced the SBM subsequently abandoned it because of the mostly unfavorable outcomes on balance. For example, the United Kingdom abolished its mandatory competitive pool—the most advanced form of the SBM. Ukraine, which adopted a United Kingdom–style, single-buyer power pool in the 1990s, amended the Electricity Law in 2013 to allow the full-scale liberalization of the wholesale electricity market by gradually transitioning from the existing and poorly functioning SBM to a bilateral contract market, day-ahead spot market, balancing market, and ancillary services market.

Regional Electricity Links: From Net Imports to Net Exports

As part of the former Soviet Union, Kazakhstan was a member of the integrated Central Asia Power System (CAPS) together with the Kyrgyz Republic, Uzbekistan, Tajikistan, and Turkmenistan. Water and electricity were inextricably linked under CAPS. For Soviet central planners, the key rationale behind CAPS was water sharing, not electricity exchange. The primary goal of regulating the flow of the Amu Darya and Syr Darya Rivers in upstream countries (the Kyrgyz Republic and Tajikistan) was to provide a reliable water supply for agriculture in downstream countries (Kazakhstan and Uzbekistan) during the irrigation season. Generation of electricity at upstream hydropower plants played a secondary role. Electricity was generated mostly during the irrigation season when large volumes of water were released. The provision of compensating energy supplies (fuels and electricity) was centrally arranged by Moscow to upstream countries to allow them to accumulate river flow in reservoirs during fall and winter. This was a centrally orchestrated system of quota allocations for the five member countries, covering water and the compensating fuel supplies as an integrated regional system. Water and fuels were exchanged between the republics as free and shared goods.

The system worked reasonably well as long as the five countries were part of the Soviet Union at the time. The system began to fray at the edges after 1991, however, as the newly independent countries began to assert competing national interests in the absence of Moscow's coercion. Ad hoc bilateral negotiations and agreements to exchange water for fuel regularly broke down, with each of the participants essentially undermining the basic terms of the agreements. There was no explicit recognition of the obligation of downstream states to pay for the annual and multiyear water storage services that upstream countries provided at considerable economic cost. Interstate, geopolitical rivalries—particularly between Kazakhstan and Uzbekistan—a long history of distrust, and lack of political cooperation compounded matters further. Of the five original members, today only Kazakhstan, the Kyrgyz Republic, and Uzbekistan remain synchronously interconnected. The operational coordination practices, applied in the region earlier, changed following the increasing level of disintegration. CAPS's Tashkent-based and technically outdated United Dispatch Center's role switched from direct operational control and subordination (as in former Soviet times) to a largely advisory role (including data exchange) to the national dispatch centers within CAPS. The recently very limited intra-CAPS power flows are not regionally dispatched. No mutually accepted rules exist for the regulation of cross-border power exchanges. For example, Uzbekistan—given its strategic location as a key transit country that borders the four other members—often demands high, above-market transit fees, thereby impeding the profitability of the established arrangements of power exchange, or arbitrarily interrupts transit deliveries across its national grid.

Coordination of the Kazakhstani power system with the operation of the Russian and Central Asian systems is based on simple monthly net flow agreements with Russia and agreed hourly transfer schedules with the Central Asian partners. The scheduling issues in the Central Asian system are complex, as they are interwoven with water management issues, and they must be solved and agreed upon by all the affected countries together. Because of this complexity and the varying interests of the affected countries, as well as actual weather developments, the conclusion of the annual agreements governing the operations of Central Asian connections is challenging and the outcomes may be uncertain.

Within CAPS, the Kazakhstani-Uzbek bilateral electricity trade is the most troubled bilateral relationship, vividly reflecting the serious shortcomings in operational coordination and in ensuring system security within CAPS. Instead of constraining domestic power consumption during periods of peak demand, when the latter exceeds peak capacity, UzbekEnergo (Uzbekistan's national utility) is engaged from time to time in massive unscheduled power transfers from the Kazakhstani system, occasionally causing an overload of the Kazakhstani North–South interconnector, thereby activating the automatic protection control system and triggering widespread blackouts in southern Kazakhstan. Because of the unbalanced operation of the Uzbek power system, the fall and winter periods of 2011 and 2012 were particularly disruptive to the bilateral ties. For example, in September 2011—not a high-load period—the unscheduled, noncontractual Uzbek "overdraft" from Kazakhstan was 500 MW (a little less than the transfer capacity of Kazakhstan's second North–South interconnector) in the evening peak hours. This overdraft triggered an overload of the interconnector, which automatically shut down, causing wide-ranging blackouts in southern Kazakhstan that lasted for hours.[8] In 2014, unscheduled Uzbek imports from Kazakhstan totaled 652 million kilowatts/hour, or nearly 1 percent of total generation.

Following difficult negotiations with KEGOC over the terms of the payment (including the price of "overdrawn" electricity)—even when UzbekEnergo agreed to pay belatedly—these events have caused serious damage to regional cooperation. If it is repeated, this type of situation may push Kazakhstan to a formal disconnection from Uzbekistan, which would be the end of CAPS as it is known. In recent years, KEGOC and the Kazakhstani authorities have raised the possibility of permanent disconnection of the two power systems if Uzbekistan continued the practice of large-scale unscheduled power transfers, thus violating the accepted and legally binding norms of simultaneous, interconnected operation.[9] The serious anomalies within CAPS and the related interstate conflicts have only strengthened the government of Kazakhstan's desire for total electricity independence.

Unlike Uzbekistan, Kazakhstan may leave CAPS without suffering prohibitively large damage in system stability and reliability. Overall, the national grid is in good technical shape and reasonably well-connected internally in the crucial north–south direction, and generation capacity has begun to increase in recent years. Most important, strong and multiple transmission connections to Russia—amounting to nearly 11,000 MW—ultimately ensure the operational stability and security of the Kazakhstani system. With the planned construction of the third North–South interconnector, Kazakhstan is set to further strengthen its system by tapping the vast renewable (especially hydro and wind) potential in the northern and eastern parts of the country. This will reduce the country's partial dependence on the Kyrgyz Republic for frequency regulation and power deliveries during winter peak load.

As a result of long-running disintegration, mutual electricity trade within CAPS has collapsed to a trickle—from 25 terawatt hours (TWh) in 1990 to 2–3 TWh in recent years. The tenfold decrease points to the fact that, within CAPS, trade creation was replaced by trade aversion under the pursuit of energy independence and trade diversion toward new extra-CAPS partners. The unified CAPS is slowly disappearing. As table 4.1 illustrates, Kazakhstani electricity imports from CAPS partners (mostly the Kyrgyz Republic and Uzbekistan) fell sharply—from nearly 9 TWh in 1990 to an insignificant level in 2014. Imports from Russia also fell dramatically. Under the government of Kazakhstan's long-term energy independence concept, no net electricity imports are projected for 2018–25 (Government of Kazakhstan 2014).

Kazakhstan does not appear interested in revitalizing CAPS in its original five-country setting, claiming that CAPS has been overtaken by new developments in the region and beyond; therefore, it no longer provides an effective framework for the promotion of electricity exchange in the region. Kazakhstan is actively seeking ways to achieve and maintain full electricity independence—a key objective of the Energy Concept 2030. As table 4.1 shows, since 2013 the country has been a net electricity exporter, with about 3 TWh exported annually in the past two years, predominantly to Russia. The government of Kazakhstan plans to

Table 4.1 Kazakhstan's Electricity Trade, 1990–2014
GWh

	1990	2000	2010	2011	2012	2013	2014
Imports	**17,336**	**2,957**	**2,035**	**3,741**	**2,568**	**884**	**644**
Russian Federation	7,590	1,848	401	1,284	1,184	510	560
Central Asia	9,746	1,109	1,634	2,457	1,384	374	84
Exports	**587**	**0**	**564**	**1,808**	**1,371**	**3,216**	**2,918**
Russian Federation	0	0	564	1,490	913	2,811	2,064
Central Asia	0	0	0	318	458	405	854
Net flow (exports–imports)	**−17,336**	**−2,957**	**−1,471**	**−1,933**	**−1,197**	**2,332**	**2,274**
Russian Federation	−7,590	−1,848	163	206	−271	2,301	1,504
Central Asia	−9,746	−1,109	−1,634	−2,139	−1,197	31	770

Sources: KOREM; for 1990, KEGOC.

modernize existing power facilities and construct new power plants, not only to meet internal demand reliably, but also to increase the country's power export potential. These plans are based on the country's abundant availability of low-cost coal and the much higher export prices compared with domestic price levels, which the government—under the Energy Concept 2030—will allow to increase only moderately in the longer term. The Energy Concept 2030 envisages a considerable net export surplus of 1,500–1,900 MW or about 11–13 TWh—by the end of the current decade, a rather aspirational, if not overly ambitious, goal (Government of Kazakhstan 2014). More recently, Kazakhstani officials estimated the country's power export potential at up to 2,100 MW (or 15 TWh)—about five times higher than the export level attained in 2013–14.[10]

Like its CAPS partners, Kazakhstan actively seeks opportunities to increase its electricity exports, primarily outside Central Asia, targeting Afghanistan, Belarus, China, Pakistan, and Russia as main destinations for surplus electricity. There are also plans to increase exports to the Central Asian countries that face sustained power deficits, such as the Kyrgyz Republic and Uzbekistan. Kazakhstan has expressed an interest in the CASA-1000 program, which envisages electricity transmission from Central Asian countries (primarily the Kyrgyz Republic and Tajikistan) to South Asia (Afghanistan and Pakistan). The much higher electricity prices prevailing in South Asia, compared with domestic and Russian tariffs, have created a strong incentive for Kazakhstani power exporters to link up with the CASA-1000 program. Once it is completed toward the end of this decade, the third North–South interconnector—running relatively near the Kazakhstani-Chinese border in eastern Kazakhstan—will open a possible new "energy bridge" to the power-hungry western Chinese markets.[11] There are also plans to establish a unified regional electricity market in the framework of the Eurasian Economic Union and Eurasian Customs Union, involving Armenia, Belarus, Kazakhstan, the Kyrgyz Republic, and Russia. This market would allow barrier-free exports of cheap Kazakhstani electricity to these important markets. However, the common electricity market has yet to be designed.

High Energy Intensity

Kazakhstan ranks among the top 10 most energy-intensive economies in the world. It uses three times as much energy per unit of gross domestic product (GDP) (based on purchasing power parity) compared with the average for the Organisation for Economic Co-operation and Development (OECD) (figure 4.5). Mirroring the high energy intensity, Kazakhstan is the fourth-most carbon-intensive country in the world. With 1.4 kilos of carbon dioxide per U.S. dollar of GDP emitted in 2008, Kazakhstan is more than twice as carbon intensive as the Europe and Central Asia regions, on average, and more than three times as intensive as the OECD average.

Energy is used very inefficiently in Kazakhstan, reflecting the legacy of the Soviet era. The economy is highly energy intensive and dominated by extractive industries and low-level commodity processing. Moreover, dated and inefficient infrastructure, low energy prices mirroring the country's rich fossil fuel endowments and distorted pricing, and the lack of targeted energy-efficient policies and an enabling institutional framework contribute to the inefficient use of energy. The publicly released data show that the energy intensity of GDP decreased by 18.6 percent in 2013, exceeding the 2015 target (figure 4.6). However, the latest data from the Statistics Committee show that the energy intensity of GDP went up significantly in 2014, only a 1.7 percent reduction compared with the base year of 2008. The high energy intensity of GDP in Kazakhstan is mainly caused by (a) the high contribution of energy-intensive

Figure 4.5 Energy Use per US$1,000 Gross Domestic Product, Selected Countries, 2009

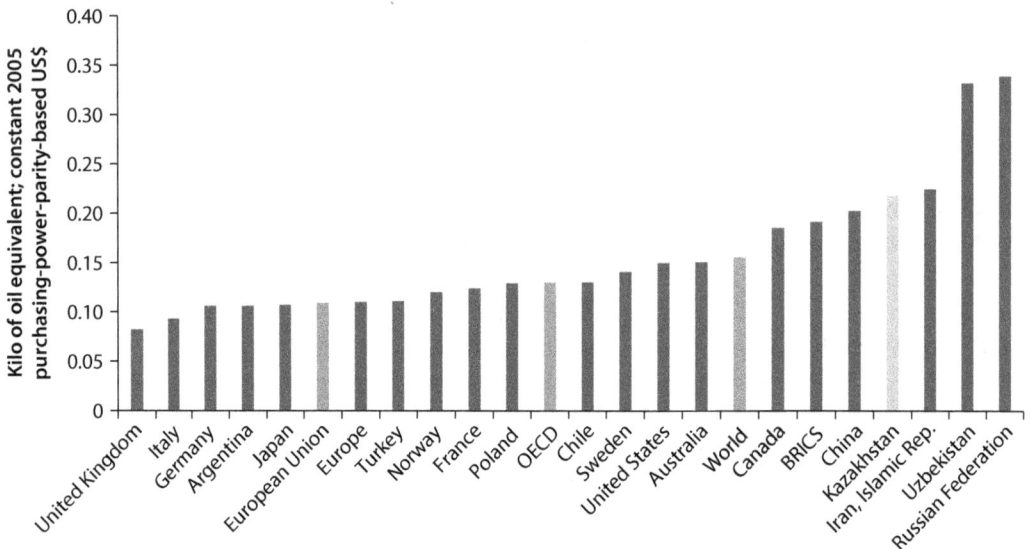

Note: BRICS = Brazil, the Russian Federation, India, China, and South Africa; OECD = Organisation for Economic Co-operation and Development.

Figure 4.6 Energy Intensity of Gross Domestic Product in Kazakhstan, 2005–14

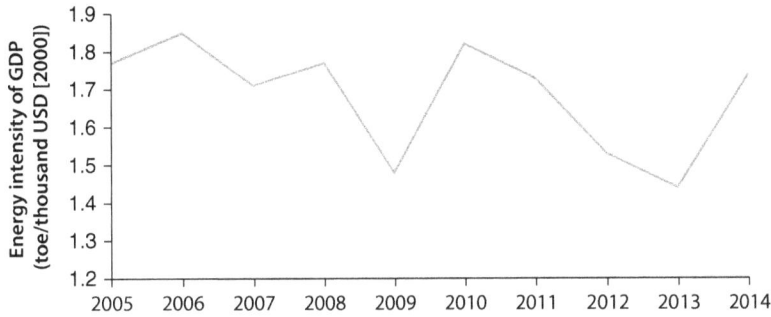

Source: Kazakhstan Statistics Committee.
Note: GDP = gross domestic product; toe = tons oil equivalent.

industrials, including energy and the extractives sector, to GDP; (b) the low energy efficiency level in key energy-consuming sectors; and (c) adverse climate conditions. The factors contributing to the increase of energy intensity in 2014 were mainly caused by the high dependence of Kazakhstan's GDP on oil resources and the global slump in oil prices.

The high energy intensity confers significant costs on the country in economic competitiveness, public health, and the environment. International comparisons show that Kazakhstan's industrial sector is significantly more energy intensive than that of most countries. This negatively affects the competitive position of Kazakhstani semi-manufactured goods on international markets, especially in the energy-intensive metal product categories. Inefficient use of electricity contributes to power shortages, especially amid tightening supply and demand balance, and adversely impacts regional economic development and social welfare. Energy-related pollution is one reason for the existence of several environmental "hot spots" in the country, with localized pollutants (such as mono-nitrogen oxides, sulfur oxides, and particulate emissions) posing significant health risks.

Historically, energy efficiency was not a high priority of the government of Kazakhstan. A Law on Energy Saving was adopted in 1997, but has remained mainly declarative in nature, because of the lack of specific national goals on energy efficiency improvements and implementable action plans. Recently, the government has devoted increasing attention to energy efficiency as a policy priority to prevent serious growth-slowing energy shortages, improve industrial competitiveness and environmental performance, and mitigate the social consequences of the recent rapid rise in domestic energy prices. In March 2010, the President of Kazakhstan set the goal to reduce the national economy's energy intensity by 10 percent by 2015 and 25 percent by 2020 (from 2008 levels).

Specific Barriers to Energy Efficiency Investments

The abundance of opportunities for profitable energy efficiency investments is in sharp contrast with the limited number of successful energy efficiency projects and the low volume of actual investments, particularly in the public sector. Reasons for this disparity are informational, technical, financial, institutional, and policy/procedural barriers that constrain the promotion and market penetration of energy efficiency. These include the following:

- *Energy pricing.* Energy tariffs determine the financial viability of energy effi- ciency investments. Despite substantial recent increases, government-regulated retail electricity and heat tariffs are still considered significantly below full cost-recovery levels. Furthermore, in most cases, heat services are billed based on regulated norms rather than consumption, which does not encourage energy savings by end users. This is a major factor that reduces the financial viability of energy efficiency projects.
- *Financial barriers.* The shortage of readily available and affordable debt financ- ing and/or sound energy efficiency financial mechanism(s) in place represents a key barrier to the uptake of energy efficiency projects in public facilities. Commercial banks are generally not familiar with the financial and technical issues involved in energy efficiency projects, and the banks perceive the risks to lending to municipal and other public entities—as well as the transaction costs of such projects—to be high. The excessively risk-averse bank behavior, high collateral requirements, and lack of viable delivery mechanisms also constrain financing for energy efficiency. On the one hand—as with many post-Soviet states—a culture of municipal financing and credit is lacking, with many public entities reliant on state budget transfers to cover most, if not all, of their expenses, and they face borrowing restrictions. On the other hand, the state budget for energy efficiency funding for municipal and public enti- ties is potentially available, but it requires the development of financing frameworks.
- *Lack of information and weak technical capacities.* The lack of technical skills, information, and awareness hampers the demand for energy efficient prod- ucts and services. Frequently, potential project sponsors lack the capacity to develop high-quality, bankable energy efficiency investment proposals, or are skeptical of the actual energy cost savings. Therefore, end users, particularly those in the public sector, are reluctant to undertake investments if they can- not be sure the operational savings will pay for the underlying investments. The energy efficiency market is currently underdeveloped because of weak technical capacity and lack of demand for energy efficient services and goods. For instance, there are several energy audit companies, local and subsidiaries of international companies, but almost no energy service companies operating in the market.
- *Institutional and regulatory barriers.* Despite the government's recent pol- icy efforts, the institutional and regulatory framework for energy effi- ciency remains largely fragmented, and most measures have yet to be fully

implemented. Although a new energy efficiency law was adopted, secondary legislation and regulations still need to be developed and enforced, including budgeting, procurement, certification schemes, auditing, and benchmarking. Furthermore, the public sector suffers from a range of procedural barriers, from budgeting to procurement, which tend to be rigid in nature and prevent many energy efficiency improvements from being effected.

Notes

1. Including, inter alia, "Amendments to the Electricity Law" (effective May 5, 2015) and "100 Concrete Steps Set Out by President Nursultan Nazarbayev to Implement Five Institutional Reforms," May 20, 2015.

2. According to the amended Electricity Law, price caps will apply for the Capacity Market and the energy market. The price caps will be differentiated by groups of power plants. It is anticipated that different groups will be created for the capacity price caps compared with the energy price caps. This means, for example, that two power plants may be in the same group for the capacity price caps, but in different groups for the energy price caps.

3. These agreements fix the tariff, amount, and duration of capacity supply.

4. On May 5, 2015, President Nazarbayev noted, "The electricity sector has a large share of outdated equipment and infrastructure. Problems are aggravated by a nontransparent system of tariffs. In order to effectively address these issues, a SBM will be introduced. The centralized purchase of electricity will allow us to reduce tariff differences between our regions and contain the rate of tariff increase." A uniform wholesale tariff is not a sound, economically justified goal for a large country, such as Kazakhstan, which has enormous regional differences in several respects. Tariffs across regions should reflect the significant underlying cost differences in the delivered energy. For example, there is no economic justification for an identical wholesale tariff for the energy-surplus Ekibastuz region in the North and the energy-deficit Almaty region in the South.

5. For example, under some of the proposals favored by some Kazakhstani experts, centralization of all electricity trades would take place under a compulsory electricity auction in which only the generators would submit price bids for different time periods (from day-ahead to year-ahead) against nationally aggregated demand. International experiences with such administratively complicated "gross pools" are not positive.

6. Examples are the easier system-level balancing of supply and demand, unified wholesale price (which simplifies tariff regulation), shielding generation financiers from market and regulatory risk, and preservation of the sector ministry's authority on major investments in generation capacity and the sector as a whole.

7. For example, this is the case in the Republic of Korea, where the national power company, KEPCO, incurs permanent losses as a single buyer because of the below-cost, regulated selling price of electricity.

8. An even larger systemwide calamity was caused by Uzbekistan in January 2011, when a 300 MW Uzbek unscheduled "overdraft" led to the overload of two major Kyrgyz transmission lines connected to a cascade of hydropower plants, which had to be shut down. The "overdraft" resulted in the cumulative closure of a massive 800 MW capacity in the synchronously integrated CAPS system and, inter alia, led to the automatic

emergency shutdown of the Kazakhstani North–South interconnector three times, causing blackouts for hours.

9. Given the Uzbek generation system's failure to meet winter peak demand and the lack of direct transmission links to Russia—the ultimate system stabilizer of the region— Uzbekistan would suffer disproportionately more from Kazakhstan's possible disconnection. Therefore, Uzbekistan must be interested in finding a sound, lasting solution for unscheduled power imports, barring drastic and costly domestic demand curtailments. One possible alternative to consider is to channel the country's unplanned electricity demand into an organized short-term market, such as the already functioning Kazakhstani spot market, KOREM, which is open to accommodate foreign traders, including UzbekEnergo, as registered market participants. Rather than resorting to disruptive, unscheduled power transfers (that is, imports), UzbekEnergo would submit to KOREM's Internet-based trading floor demand bids on a quarterly, monthly, weekly, day-ahead, or intra-day basis. With Kazakhstan's growing generation capacity and the possibility of Russian imports via Kazakhstan, KOREM could competitively seek out suppliers to match the Uzbek demand bids. Because of the long-running rivalry between the two nations, the Uzbek side may not find this arrangement politically acceptable. A politically more palatable option is to create an organized short-term power exchange in Uzbekistan and couple it with KOREM. Such market coupling between national power exchanges is widely and successfully used in Europe, where most countries, particularly small ones, operate a relatively low-liquidity national power exchange. The main purpose of market coupling is to maximize the total economic surplus of the participants: cheaper electricity in one country can meet demand and reduce prices in another country. Prices will tend to equalize across adjacent countries where there is sufficient cross-border transmission capacity, such as in the case of Kazakhstan and Uzbekistan.

10. See htpp://www.primeminister.kz/news/show/24/Kazakhstan-gotov-eksportirovat -bolee-2-tys-mvt-elektroenergii-a-bekenov/02-04-2015.

11. In China, the wholesale electricity price is five to ten times higher than in the Ekibastuz generation hub. Construction of a transmission line from Ekibastuz to Üto (China) has been under consideration for some time.

References

GoK (Government of Kazakhstan). 2014. "Energy Concept 2030." Concept on Development of the Fuel and Energy Complex of Kazakhstan until 2030.

Samruk-Energy. 2014. "2014 Annual Report of Samruk-Energy."

CHAPTER 5

Supply-and-Demand Balance and Least-Cost Analysis

Least-Cost Planning Study: Objective and Approach

The system modeling analysis provides an updated and refined view of Kazakhstan's capacity and generation mix for 2015–45, and informs decisions on the selection of various alternative generation technologies and their sizing and sequencing. A long-term, least-cost investment study used Power System Research (PSR) planning software (see appendix B for a description of least-cost expansion modeling), operated by the Power Systems Planning team of the World Bank's Energy and Extractives Global Practice. The PSR software was developed by PSR Inc., a global provider of technological solutions and consulting services in the areas of electricity and natural gas since 1987. The analysis uses data from earlier studies with supply and demand projections and sector information provided to create a mathematical power system model, updated with recent actual numbers and updated capital expenditure (CAPEX) and operational expenditure (OPEX) cost estimates. This chapter presents a quantitative analysis of

- *Current detailed power supply and demand balance* that relates to utilization, availability of existing generators, and past demand trends;
- *Least-cost expansion plan* of generation and transmission capacity over 2015–30, to identify the most economic set of new power stations, transmission lines, and transmission interconnectors, as well as their timing and associated investment requirements; and
- *Assessment of the results of the least-cost study.*

For the demand forecast, the analysis relies largely on the data used for the "2010 Roadmap for the Development of a Competitive Kazakhstan Generation Market," which was prepared by KEMA for the Ministry of Energy and Mineral Resources. The analysis includes updated information on capital costs and fuel prices. The analysis makes use of recent supply and demand data and sector information to create a mathematical model for the Kazakhstani power system. In addition, earlier planning studies and government documents are used to obtain information on projections for future supply, as well as OPEX and CAPEX estimates.

The study presents a Base Case scenario and three other scenarios that reflect alternative assumptions about demand and the implementation of specific economic and environmental policies. The result is an investment plan (2015–45) that meets demand in a least-cost manner, subject to several constraints.

Supply and Demand

The critical information around demand and supply is summarized as follows (figure 5.1):

- Power demand more or less doubled after 1999, following a "rebuilding" phase of the Kazakhstani economy (to about 94 terawatt hours [TWh] in 2014).
- Peak demand reached 13.6 gigawatts (GW) in 2014.
- Demand is projected to grow at 2.8 percent per year (to approximately 145 TWh by 2030).
- Demand growth in the Western zone is expected to be higher compared with growth in the Southern and Northern zones.
- Demand is driven mainly by industrial loads and is thus very "flat," with an associated high load factor.

Figure 5.1 Historical Generation Mix and Demand, 1990–2012

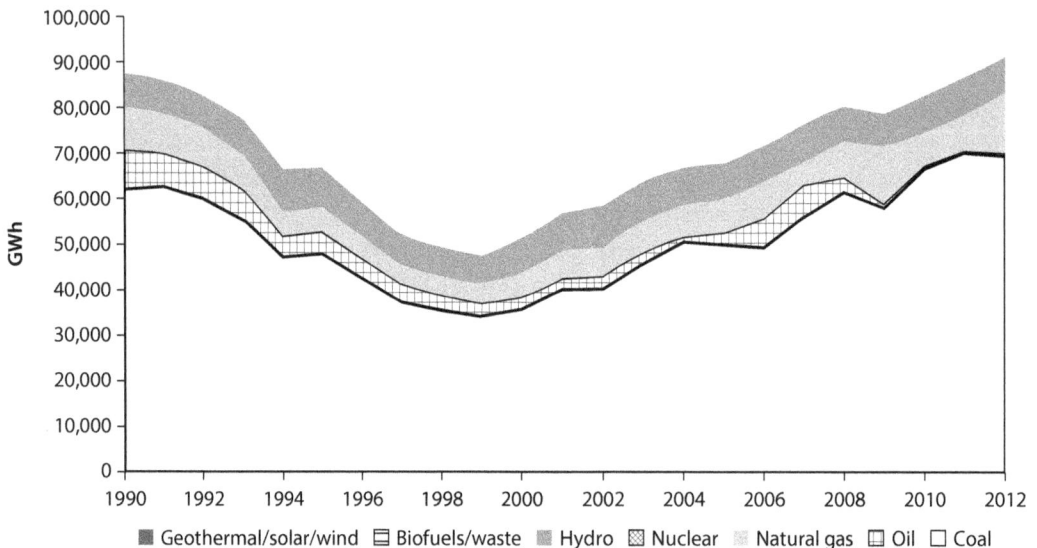

Legend: ■ Geothermal/solar/wind ⊟ Biofuels/waste ▨ Hydro ▧ Nuclear ▒ Natural gas ▱ Oil ▢ Coal

Source: IEA, available at http://www.iea.org/stats/WebGraphs/KAZAKHSTAN2.pdf.
Note: GWh = gigawatt hours.

- The current generation fleet is aged and largely obsolete. Of this, 2.2 GW of thermal generation are expected to have been decommissioned by 2030, when peak power demand is expected to have reached about 23 GW.
- Total installed capacity is 21 GW. Total available capacity is 15.2 GW.
- Hydro installed capacity is 2.2 GW.
- A tariff cap program was implemented in 2009–15 to encourage capacity rehabilitation/extension investments and help improve reliability.
- The reserve margin was 11 percent in 2014 but fell to a dangerously low 4 percent in 2012. The growth in the reserve margin after 2008 was a result of a combination of the economic crisis and the tariff cap program.
- Currently, coal accounts for approximately 75 percent of total power production.
- Fuel diversity is restricted by the currently limited natural gas network.
- Combined heat and power plants (CHPs) provide the necessary heating requirements near city centers. The current installed capacity of CHPs is 6.2 GW and constitutes 33 percent of total thermal installed capacity.

Past Demand Trend and Current Supply Mix

The annual peak electricity production increased in parallel to energy demand, growing from 8.6 GW in 2010 to 13.6 GW in 2014. The economy has been in a rebuilding phase since 2000, after a marked decline in economic growth, highlighting the significant uncertainty in demand growth. During the rebuilding phase, the annual increase in power generation and demand was more or less steady, at around 4.4 percent—or 2 TWh per year.

The high historical rate of growth of electricity consumption was driven largely by the export-oriented, electricity-intensive, heavy industry targeting mostly the Russian market. By international comparison, the share of industry in total electricity consumption is unusually high, at about two-thirds. The 10 mostly export-oriented, heavy-industry (such as mining and metallurgical) companies account for approximately half of industrial electricity consumption. The export success of these companies hinges largely on cheap coal and electricity, as well as growth of the nearby export markets.

Total electricity consumption varies significantly by region, with the Northern zone being the largest power consumer, followed by the Southern and Western zones. Peak demand for 2014 was 8,757 megawatts (MW), 2,189 MW, and 1,304 MW for the Northern, Southern, and Western zones, respectively. In all three zones, power consumption is driven mainly by "smooth" industrial loads, the reason Kazakhstan's power system experiences relatively low load variation and has a large load factor of > 70 percent (figure 5.2).

Demand Projections

Projections about future demand are regularly published by the government of Kazakhstan and the Kazakhstan Electricity Grid Operating Company (KEGOC). Projections are also available in studies prepared by KEMA, the Asian Development Bank (Fichtner GmbH & Co. 2012),[1] and the Energy Association

Figure 5.2 Hourly Load per Zone on November 28, 2014

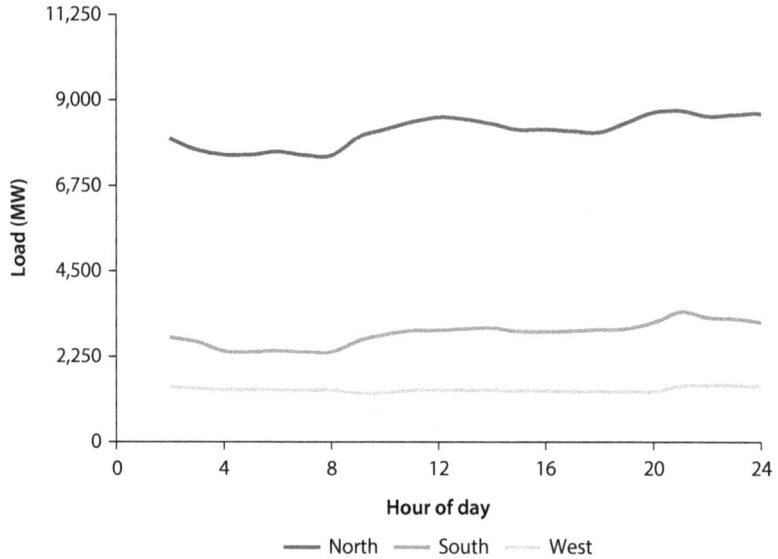

Source: Kazakhstan Electricity Grid Operating Company.
Note: November 28, 2014, was the day with the third-highest annual peak for the year. GWh = gigawatt hours.

of Kazakhstan (KazEnergy). Econometric models[2] that use historical data to predict demand as a function of gross domestic product (GDP), population, industrial or general economic development, and other demand drivers are used for the forecasting.

The demand study carried out by KEMA ("2010 Roadmap for the Development of a Competitive Generation Market") projects moderate, linear demand growth of about 2.8 percent. The KEMA 2010 study adjusted its initial forecasted demand growth of 4 percent to match KEGOC's forecasted demand, which assumes the past growth trend will decrease in the future (figure 5.3).

KEMA's road map (figure 5.4) assumes high- and low-demand growth scenarios (4.2 and 1.4 percent growth, respectively) to account for higher and lower industrial and GDP growth.[3] As figure 5.3 shows, demand growth is not the same in each zone. Using 2015 as the reference year, growth is predicted to be highest in the Western zone (4.15 percent) and lowest in the Northern zone (1.8 percent), with demand growth in the Southern zone at 3.35 percent. As a result, the Kazakhstani peak load is expected to reach about 23 GW by 2030.

Hydro and Renewable Energy
Hydro
There are currently six large hydropower plants with derated capacity > 100 MW that total 2,160 MW, and many smaller plants that add another 96 MW (for a total of approximately 2,255 MW). The hydro fleet, similar to the thermal fleet, is largely from the former Soviet period, with considerable capacity lost because of aging.

Figure 5.3 Energy Demand and Peak Load Projections by the Kazakhstan Electricity Grid Operating Company, to 2030

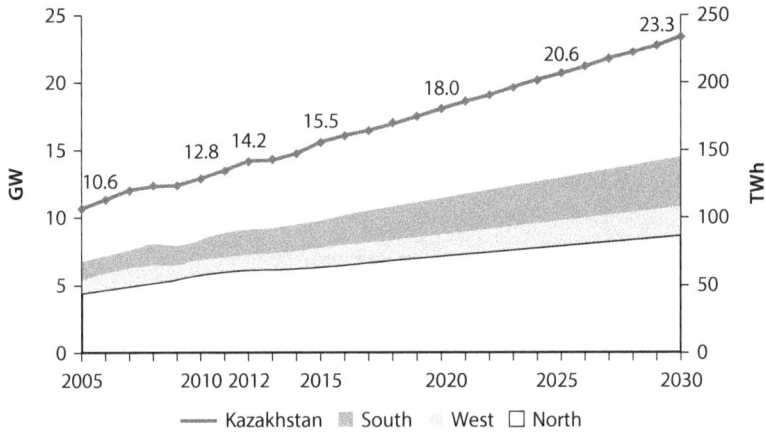

Source: KazEnergy 2013.
Note: GW = gigawatts; TWh = terawatt hours.

Figure 5.4 Base, High-, and Low-Demand Scenarios, 2001–25

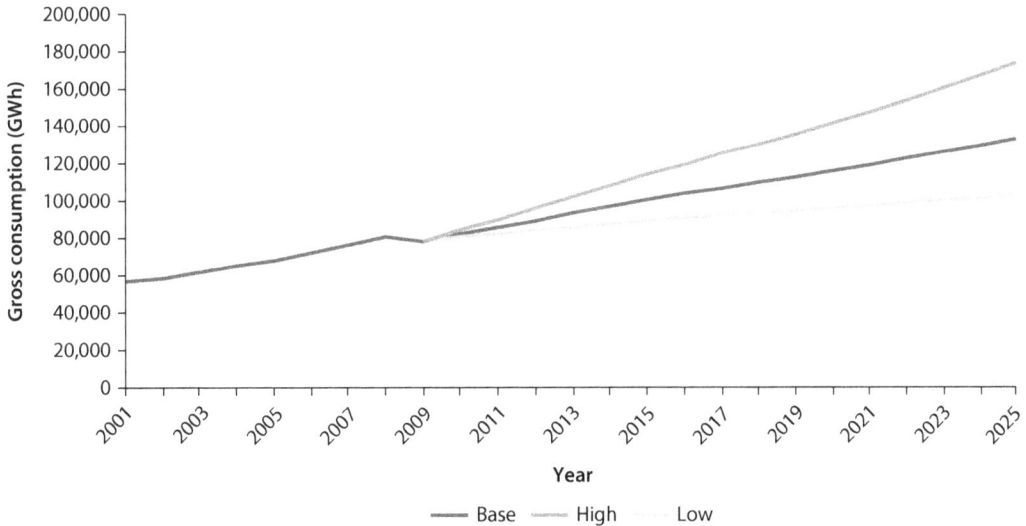

Source: KEMA 2010.
Note: GWh = gigawatt hours.

The government of Kazakhstan plans to rehabilitate existing hydropower plants and add 302 MW by 2020, in addition to the current available capacity.

Wind and Solar Power

According to a 2011 KazEnergy report, an action plan to commission 793 MW of wind power and 77 MW of solar power by 2020 has been approved. However, the government's renewable energy targets are more aggressive now; current targets for solar and wind power penetration are 3 percent by 2020 and 10 percent by 2030 (DNV-GL 2015a). Additional targets include 30 percent alternative energy[4] by 2030 and 50 percent by 2050 (KazEnergy 2013). As of now, under construction (or completed) there are about 500 MW of wind power and 300 MW of solar power, scheduled to be on line by 2020. The largest solar plant under construction is the 100 MW Cogenhan project in Jambyl province (Southern zone), and the largest wind power project (200 MW) is in the Zhambyl region (Southern zone).

Current Capacity Expansion Plan: 2015–30

Several attempts have been made to develop a comprehensive system expansion plan—the latest one by KazEnergy in 2013 (KazEnergy 2013). The KazEnergy plan predicts a steep increase in coal generation, from 60 TWh per year at present to more than 100 TWh per year by 2030, which would thus remain the bedrock of the Kazakhstani power system. Gas-based generation will significantly expand, however, and the addition of wind and solar power by 2030 will diversify the mix to some extent (figure 5.5).

To meet this expansion, KazEnergy's *National Energy Report 2013* reflects an investment requirement of approximately US$54 billion over 2013–30

Figure 5.5 Projected Generation Mix, 2012–30

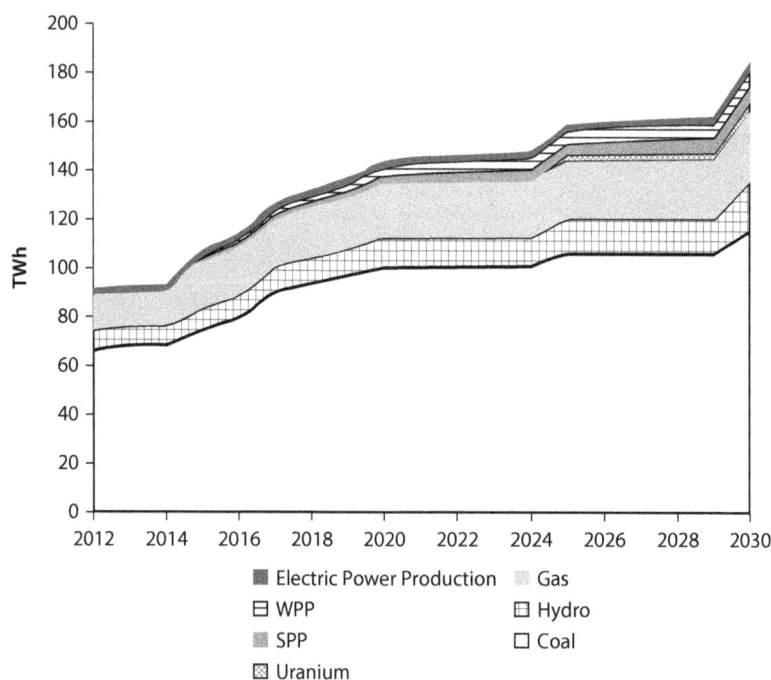

Legend: Electric Power Production; Gas; WPP; Hydro; SPP; Coal; Uranium

Source: KazEnergy 2013.
Note: SPP = Solar Power Plant; TWh = terawatt hours; WPP = Wind Power Plant.

Table 5.1 Investment Requirements
US$ millions

Line designation	2013–15	2016–20	2021–25	2026–30	Total for 2013–30
Investment, including in	8,611	19,485	11,217	14,681	53,993
Technical upgrading	2,547	2,112	417	417	5,196
Extension	1,934	4,407	275	55	6,670
New commissionings	4,130	12,966	10,525	14,505	42,127

Source: KazEnergy 2013.

Table 5.2 Investment, by Plant Type
US$ billions

Power plant type	Exaggerated estimation of investments by 2030 (2013 prices)
Coal-fired	23.5
Gas-fired	4.6
Nuclear	5.5
Hydro (inclusive of midget ones)	5.4
Wind	6.9
Solar	8.1
Total	54.0

Source: KazEnergy 2013.

(table 5.1), or more than US$3 billion per year. This investment will (a) upgrade and extend the existing capacity (about US$12 billion, the bulk of which occurs before 2020); and (b) build new capacity worth US$42 billion, including US$25 billion in new investments following 2020.

Additional capacity, planned over the next 15 years, is close to 18 GW, including 7.5 GW of coal (including refurbished capacity), 1.8 GW of gas, 0.9 GW of nuclear, 2.7 GW of wind, and 2.3 GW of solar. Table 5.2 shows the investments by type of plant—coal alone accounts for US$23.5 billion, and coal and gas, together, account for half of the total investment of US$54 billion.

Least-Cost Planning Analysis: 2015–45

The analysis modeled four scenarios. These are described in detail in the following subsections and summarized in table 5.3.

Base Case Scenario

As the most likely scenario, the Base Case optimizes generation and transmission and takes into account the policies, goals, and investment projects already in the process of being implemented or very likely to be implemented. The Base Case scenario optimizes expansion of generation and transmission, considering

Table 5.3 Summary of the Basic Assumptions for Each Scenario

Scenario	Base Case	Least-Cost Case	Green Case	Regional Export Case
Demand	• 2014 Demand/ peak demand (94 TWh/13.6 GW) • 2.8% annual growth	Same as Base Case scenario	• Significant reduction in demand as a result of energy efficiency measures • 2.3% growth up to 2030; • 1.2% growth up to 2045	• Internal demand same as Base Case scenario • 13 TWh of electricity must be exported to Russia and Central Asia by 2030 • 13 TWh exported annually after 2030
Fuel supply options	• All fossil fuels are domestic • Coal, natural gas, oil, uranium • Coal is not available in the West • Gas in not available in the North	Same as Base Case scenario with the addition that partial gasification in the North to convert some CHPs from coal to gas is subject to optimization	Partial gasification of the North zone, starting in 2020, to convert CHP generation equivalent of 4bcma from coal to gas; rest of supply options same as Base Case scenario	Same as Base Case scenario
Fuel costs	• Natural gas: West $0.059/m³; South $0.097/m³ • Coal: North $15/ ton; South $22/ton • Oil $52.5/BOE; Nuclear $0.4/GJ	Same as Base Case scenario with the addition that natural gas in the North for conversion of CH P power plants costs $0.082/m³ ($0.09/m³ in 2020).	Same as Base Case scenario with the addition that natural gas in the North for conversion of CHP power plants costs $0.082.m³ ($0.09. m³ in 2020).	Same as Base Case scenario
Thermal supply options	• Coal: CHP/ Supercritical • Gas: CCGT/CHP/ OCGT • Oil: Steam • Nuclear	Same as Base Case scenario	Same as Base Case scenario	Same as Base Case scenario
VRE targets	• 3% of solar and wind by 2020 • 10% of solar and wind by 2030 • VRE growth trend is assumed to continue up to 2045	VRE projects are subject to least-cost optimization	The installed capacity of VRE is left same as Base Case scenario even though demand is lower	Same as Base Case scenario
Other targets	• 1 GW of nuclear by 2030		• 30% of VRE + nuclear by 2030; 50% by 2045 • 15% CO_2 reduction of 2012 levels by 2030; 40% by 2045	Same as Base Case scenario

table continues next page

Table 5.3 Summary of the Basic Assumptions for Each Scenario (continued)

Scenario	Base Case	Least-Cost Case	Green Case	Regional Export Case
Regional interconnections	• North-South transmission corridor of 1,350 MW and the plan to increase it to 2,100 MW in 2018 are modelled • West is modelled as island zone • No regional interconnections modeled	Same as Base Case scenario with some additions: Possible addition to the already planned interzonal transmission capacity including West/North and West/South interconnections are subject to optimization	Same as Base Case scenario	• Kazakhstan is interconnected with Russia through a regional interconnection with capacity of 10,590 MW • Kazakhstan is connected with Kyrgyz Rep. and Uzbekistan through 2,460 and 940 MW interconnections, respectively • Above transmission capacities are assumed to continue in the future

Note: The fuel price includes the cost of building a gas network in the Northern zone. bcma = billion cubic meters per annum; BOE = barrel of oil equivalent; CCGT = combined cycle gas turbine; CHP = combined heat and power plant; CO_2 = carbon dioxide; GJ = gigajoule; GW = gigawatts; m^3 = square meters; MW = megawatts; OCGT = open cycle gas turbine; TWh = terawatts; VRE = variable renewable energy.

demand projections and the availability and cost projections for various fuels. The Base Case scenario assumes the following:

- Demand growth projected at 2.8 percent per year (aligned with the KEGOC forecast).
- The government's aims to achieve 3 percent of variable renewable energy (photovoltaic [PV] and wind) penetration by 2020 and 10 percent by 2030.[5] The analysis also includes information on installed capacities, by technology and region, from a recent variable renewable energy integration study (DNV-GL 2015a).
- The government's plan for the third North–South transmission line, which will increase transmission capacity for the North–South corridor from 1,350 to 2,100 MW.
- The new coal-fired supercritical Balkhash power plant, with installed capacity of 1,320 MW.
- The extension of the Regional State Power Station, GRES-2 (by 525 MW) at Ekibastuz.
- The government's plans to develop capacity for nuclear technology and bring on line 1,000 MW of nuclear power by 2030.
- KEGOC's Master Plan for rehabilitation/extension/decommissioning of existing generators, as presented in the Green Economy Concept.

Green Case Scenario

The Green Case scenario was designed to model a transition toward green growth. Most assumptions are adopted from the Green Economy Concept, the implementation of which appears to have stalled. This scenario aims to identify economic benefits/costs in the power sector of an economywide energy efficiency program that aggressively reduces the growth of demand (and peak demand). In addition to the Base Case scenario assumptions, the Green Case scenario includes the following:

- A target to decrease annual carbon dioxide (CO_2) emissions by 40 percent by 2050 from 2012 levels.
- A target to achieve 50 percent penetration of carbon-free technologies in the energy mix (hydro, PV, wind, and nuclear) by 2050.

Regional Export Case Scenario

This scenario provides insights on economic benefits and costs if Kazakhstan were to invest in additional capacity to increase gradually its export capability. This plan supports Kazakhstan's efforts to achieve and maintain full electricity independence, reflected as an objective in the Energy Concept 2030 (Government of Kazakhstan 2014), which envisages a considerable net export surplus of about 11–13 TWh by 2030. The scenario assumes that 80 percent of total exported energy will go to the Russian Federation and Belarus (within the Eurasian Union) and the remaining part to Uzbekistan and the Kyrgyz Republic.

Least-Cost Case Scenario

This scenario optimizes the system on least-cost principles without considering any kind of mandatory government policy or target. The assumptions on demand projections, fuel costs, and supply options are the same as in the Base Case scenario. The same applies to the decommissioning plan and implementation of projects that have already been decided (such as Balkhash and GRES-2). However, variable renewable energy technologies will compete on the same footing with the rest of the technologies, and will only come on line if they reach grid parity. The same applies to nuclear power. Furthermore, conversion of CHPs from coal to gas is subject to optimization. The Least-Cost Case scenario includes transmission projects that fully interconnect and unify Kazakhstan's power system (for example, the North–West and South–West transmission projects) if proven economically justified. Finally, a variation of the Least-Cost Case scenario considers the economic cost rather than the actual cost of natural gas in the calculations.

Traditionally, the levelized cost of electricity (LCOE) is an economic assessment of the average total cost to build and operate a power-generating asset over its lifetime, divided by the total energy output of the asset over that lifetime. In this study, the "systemwide" LCOE is a similar concept that represents the average total cost to build, rehabilitate, and operate systemwide generation assets and interzonal, high-voltage transmission over the specified planning horizon, divided by the total energy output of the system over that same horizon. Therefore, in this study, "systemwide" LCOE excludes transmission and distribution but captures generation assets and the costs of the few interzonal interconnections under consideration.

Basic Assumptions

Base Case Scenario

(i)   *Demand growth.* Total demand growth of 2.8 percent. Growth is not homogenous across all zones. Projected growth is 2, 3.55, and 4.15 percent for the Northern, Southern, and Western zones, respectively (figure 5.6) (KazEnergy 2013).

(ii) *Fuel supply options.* Kazakhstan extracts domestic coal, natural gas, uranium, and oil, which are assumed to be the main fossil fuels the country will rely on for its generation mix. In addition:

- Natural gas investments are excluded in the Northern zone. Fuel supply in the region is currently constrained because of the lack of a gas network.
- Currently, no coal generation occurs within the Western zone, mainly because the great majority of gas production takes place in the region. Considering transportation costs, gas is far more economical than coal (which would need to be transported from coal mines located hundreds of miles away in the north and east of the country). Thus, coal-fired projects in the Western zone have been excluded for all scenarios.

(iii) *Fuel prices.* The prices of fuels were assumed to vary regionally:[6]

- Natural gas in the Western zone: US$1.41/gigajoule (GJ) or US$0.054/cubic meter (m^3)
- Natural gas in the Southern zone: US$2.54/GJ or US$0.097/m^3

Figure 5.6 Demand and Peak Demand Growth, 2015–45

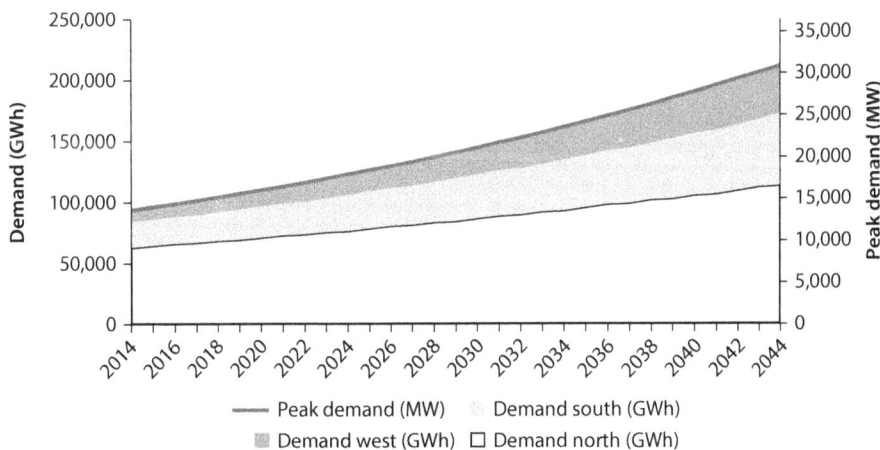

Note: GWh = gigawatt hours; MW = megawatts.

- Coal in the Northern zone: US$0.58/GJ or US$11/ton
- Coal in the Southern zone: US$1.16/GJ or US$22/ton
- Uranium: US$0.4/GJ [7]
- Oil: US$8.6/GJ or US$52.5/barrel of oil equivalent
- All fuel prices, except oil, grow at 2 percent per year; the growth rate of oil prices is 5.3 percent.

(iv) *Main characteristics and costs of generation technologies.* The technologies used in this study include the combined cycle gas turbine (CCGT), open cycle gas turbine (OCGT), CHP gas turbine, coal-fired supercritical steam cycle, coal-fired CHP, heavy-fuel oil steam cycle, fast neutron nuclear reactor, hydro, PV, and wind.[8] International Energy Agency (IEA) cost data were used mainly for Russia, because it is the closest, geographically, to Kazakhstan (noting that their markets are very different). The technology cost assumptions are given in table 5.4.

(v) *Nuclear power.* Kazakhstan has some experience in nuclear technology because it used to operate the BN-350—the first fast neutron reactor in the world. There is concern, however, about losing competence and scientific potential in the area (KazEnergy 2013). For that reason, the country is planning to build a 1,000 MW nuclear power plant by 2030 in cooperation with Russia. Because the time from project initiation to project completion for nuclear projects is long, it is assumed that a nuclear project cannot come on line before 2025.

Table 5.4 Cost Assumptions for Technologies Used in the Least-Cost Expansion Study[a]

	Efficiency[a] (%)	O&M[a] ($/MWh)	Overnight costs[a] (mil$/MW)	O&M[b] ($/kW)	Lifetime (years)
Combined heat and power plant: coal[c]	35	13.0	2.8	32	40
Coal supercritical	42	11.0	2.4	32	40
Combined cycle gas turbine	57	3.5	1.2	13	30
Open cycle gas turbine	35	7.7	0.8	12	30
Combined heat and power plant: gas[c]	35	8.8	1.4	13	30
Nuclear	41	17.0	3	104	60
Oil	45	4	1	32	20
Photovoltaic	—	0	2.1	14	25
Wind	—	0	1.7	23	25
Hydro	—	2.8	2.4	13	60

Note: The OptGen software tool does not distinguish among renewable technologies in terms of efficiency. It requires only capacity factors and installed capacities to optimize the system. For that reason, efficiency values for renewable energy technologies are not shown. kW = kilowatt; MW = megawatt; MWh = megawatt/hour; O&M = operations and maintenance.
a. Data from the International Energy Agency.
b. Data from the National Renewable Energy Laboratory.
c. For every 100 units of thermal input, it is assumed there will be 35 units of electrical output (35 percent electrical efficiency) and 40 percent useful thermal output.

Figure 5.7 Variable Renewable Energy Plan for the Optimization Period, 2015–45

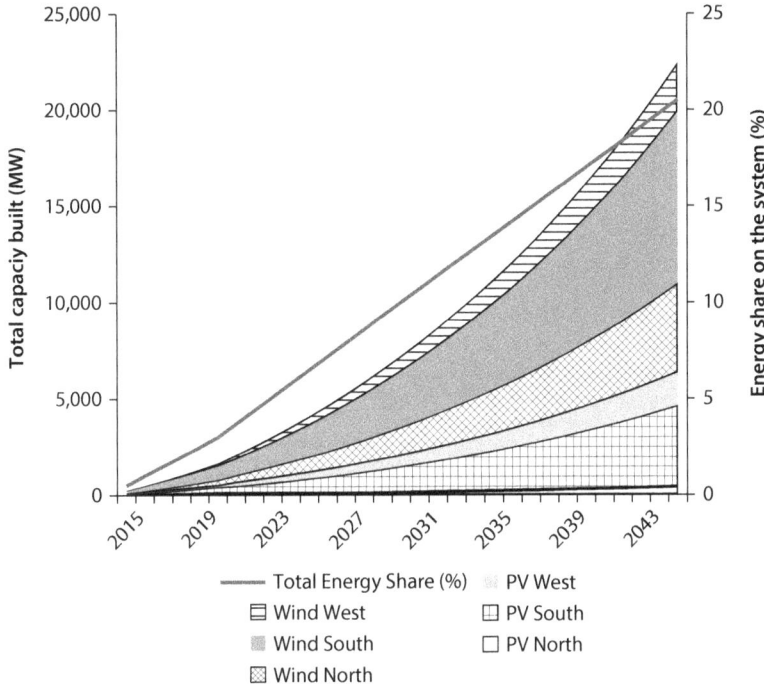

Note: MW = megawatts; PV = photovoltaic.

(vi) *Heating requirement.* To calibrate the operation of CHPs, monthly heating requirement data are used, obtained from the Ministry of Energy for 2014. The optimization process assumes that the heating energy requirement grows 1.7 percent per year. The model considers the heating requirement in the long-term plan to calculate the dispatch of CHPs and their capacity requirements. The heating requirement assumptions remain the same across all scenarios.

(vii) *Renewables.* PV and wind penetration of 3 percent by 2020 and 10 percent by 2030 is assumed, as per government policy. The trend for variable renewable energy growth in the future is assumed to continue and is used for 2030–45. The PSR software does not optimize on the basis of energy penetration targets.[9] A mix of PV and wind capacities was assumed, based on information from the DNV Area 2 study (figure 5.7) (DNV-GL 2015b). In addition, learning curves for PV and wind were applied using IEA data (IEA 2013, 2014) (figure 5.8).

The model requires average monthly capacity factors for each PV or wind invest-ment project. This calculation used hourly solar and wind data from the System

Figure 5.8 Learning Curves for Overnight Capital Costs for Photovoltaic and Wind, 2015–45

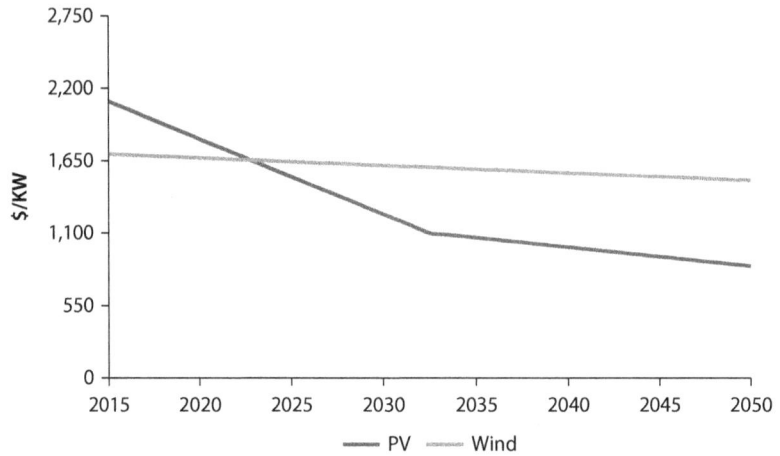

Note: MW = megawatts; PV = photovoltaic.

Table 5.5 Cost of Rehabilitation/Extension Program
(US$ millions)

	2015–20	2021–25	2026–30	2031–45
Technical upgrading	2,537	417	121	363
Extension	4,729	275	55	165

Advisor Model database of the National Renewable Energy Laboratory for each of the three electricity zones. Average annual capacity factors were calculated as follows:

- *For PV technology:* 16, 20, and 17 percent for the Northern, Southern, and Western zones, respectively.
- *For wind technology:* 30, 25, and 25 percent for the Northern, Southern, and Western zones, respectively.

(viii) *Decommissioning plan.* The current situation in the power sector is characterized by the significant obsolescence of generation and transmission assets. Kazakhstan's strategy is to modernize and extend the lifetime of the existing fleet to postpone expensive investments in new greenfield projects (table 5.5). Eventually, existing generators will decommission when their lifetime ends. The main technical activities are described in the Green Economy Concept. The original source is KEGOC's Master Plan. According to the modernization/decommissioning plan, available capacity will be gradually increased up to 2020, when the total available capacity of existing generators will be 18.9 GW. By 2030, total available capacity will have reduced by 2.2 GW (1.6 GW of coal-fired capacity, 0.6 GW of gas-fired capacity, and 0.3 GW of hydro); by 2045, only 3.1 GW of existing capacity will still be on line (figure 5.9). Decommissioning/rehabilitation constraints remain the same across all scenarios.

(ix)   *Mandatory investment projects.* Within the framework of the State Program of Accelerated Industrial-Innovative Development, the government plans to introduce 1,845 MW of generating capacity. The program includes (a) construction of the Balkhash coal-fired plant of 1,350 MW capacity, to come on line by 2022; and (b) building of a third coal-fired thermal unit in Ekibastuz (GRES-2), to be completed by 2020. These two projects are "mandatory" in the optimization model and will come on line within the above-specified dates. Assumptions about such projects are identical across all scenarios.

(x)   *Export.* Kazakhstan's power system is modeled as a three-node system (Northern, Southern, and Western), given the absence of more detailed transmission data. In the Base Case scenario, regional interconnections are not modeled because (a) it involves a very complex process to model, especially in the absence of short-term resolution trade data; and (b) regional trade has greatly reduced since 1990 and net flows are currently very low. The Western zone is assumed to be an island because it is not interconnected with the Northern or Southern zone. The North–South corridor—with a current capacity of 1,350 MW—and KEGOC's planned project to

Figure 5.9 Rehabilitation/Extension/Decommissioning Schedule of Existing Generation, 2014–45

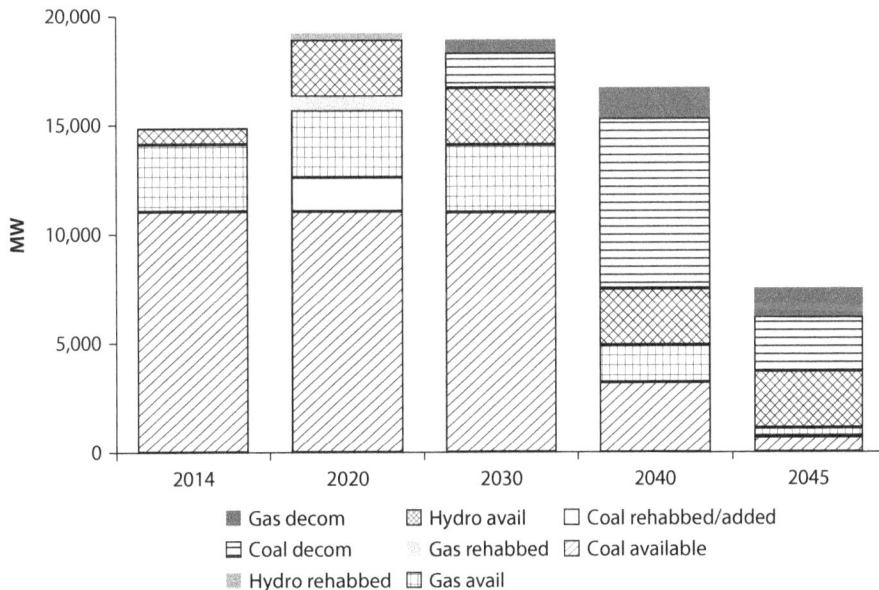

Note: MW = megawatts.

strengthen the interconnection with an additional 750 MW are modeled. The strengthening project is assumed to be completed in 2018. The Base Case scenario does not include any transmission projects; it is strictly a long-term generation expansion program. The only scenario that includes transmission expansion optimization (internal transmission projects) is the Least-Cost Case scenario.

Green Case Scenario

(i) *Demand.* In the Green Case scenario, the substantial reduction in demand growth will be based on improvements in energy efficiency on the demand and supply sides. Kazakhstan's economy is very energy-intensive; it is two to three times as intensive as the average for Organisation for Economic Co-operation and Development countries. Kazakhstani industry's carbon intensity is five times the average for the European Union. On the supply side, the average electrical efficiency of existing power plants is only 32 percent, with a potential to increase to 42–53 percent for coal-fired plants (Fichtner GmbH & Co. 2010). The Green Case scenario assumes that Kazakhstan will follow a path toward green growth, implementing energy efficiency measures that will help develop green buildings, modernize industrial equipment, and retrofit/modernize district heating systems. These measures will contribute to reduce demand growth from 2.8 percent in the Base Case scenario to 1.75 percent. The Green Case scenario assumes that demand growth up to 2030 will be 2.3 percent. After 2030, average demand growth will be reduced to 1.2 percent (figure 5.10).

Figure 5.10 Electricity Demand Growth in Kazakhstan for the Green Case Scenario, 2015–45

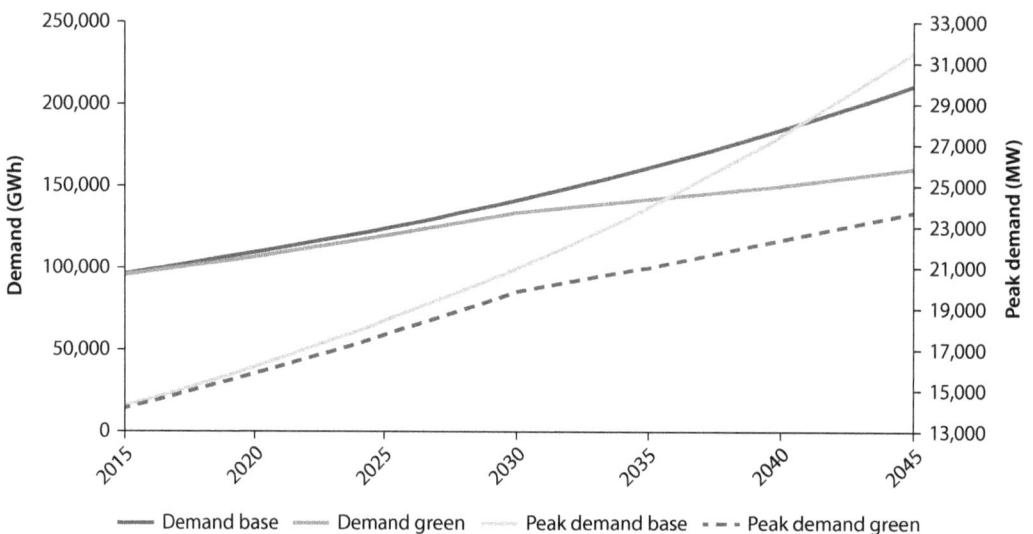

Note: GWh = gigawatt hours; MW = megawatts.

(ii) *Fuel supply options.* The Green Case scenario has the same fuel supply options as in the Base Case, with one exception. The scenario assumes that some CHP coal-fired power plants in major cities in the Northern zone will be converted to gas to help decrease air pollution. The gas availability is assumed to be 4 billion m^3 annually; this amount of gas can support about 1.5–1.8 GW of base load, gas-fired electricity production.

(iii) *Fuel prices.* Fuel prices are the same as in the Base Case scenario. In addition, the Green Case scenario assumes gas availability in the Northern zone. Gas prices in the Northern zone are assumed to be US\$3.1/GJ or US\$0.0126/m^3. The price is the sum of the cost of gas in the Southern zone (US\$2.54/GJ) plus the cost of building the required gas infrastructure (US\$0.56/GJ) (CEC 2015).[10]

(iv) *Generation technologies.* The generation technologies are the same as in the Base Case scenario.

(v) *Renewables.* In addition to the existing target of 10 percent penetration of PV and wind electricity by 2030, the Green Case scenario includes a target of 50 percent of total electricity production from carbon-free generators (that is, renewables and nuclear power). This target has to be achieved by the end of the optimization period. The assumed installed capacity of renewables is the same as in the Base Case scenario. Demand is greatly reduced, however, and as a result, solar, wind, and hydro—together— achieve 39 percent of energy penetration.

(vi) *Nuclear power.* The Green Case scenario includes a target of 2,000 MW of nuclear power by the end of the optimization period. Of this capacity, 1,000 MW has to come on line between 2025 and 2030. The additional 1,000 MW can come on line any time after that; the time of commissioning is left as an optimization decision.

(vii) *Emissions targets.* The Green Case scenario is the only one that includes an emissions reduction target. It assumes a 40 percent reduction of emissions compared with 2012 levels by the end of the optimization period. The reductions will be achieved through energy efficiency–induced demand reduction.

Regional Export Case Scenario

Demand. In the Regional Export Case scenario, the model attempts to identify the technical and economic differences from the Base Case scenario associated with assumed annual net exports of 13 TWh of electricity. This assumption is based on the fact that Kazakhstan has been a net exporter of electricity since 2013, and there is an explicit export target in the Energy Concept 2030. The annual net amount of electricity exported will gradually increase to reach 13 TWh in 2030, and will remain at that level thereafter. It is assumed that 80 percent of the exported electricity will be routed through the regional interconnections with Russia in the Northern zone (capacity of 10,590 MW); the remaining 20 percent will be routed to the Kyrgyz Republic and Uzbekistan (interconnection capacities of 2,460 and 940 MW, respectively) (KEMA 2010). The external demand is assumed to remain flat. In addition, exports to the Kyrgyz Republic and Uzbekistan are assumed to take place only during the winter period—from October to March. Figures 5.11 and 5.12 show the exported demand and required capacity over time, respectively.[11] In addition to accounting for external demand, all the other assumptions are the same as in the Base Case scenario.

Figure 5.11 Annual External Demand for Electricity, 2015–45

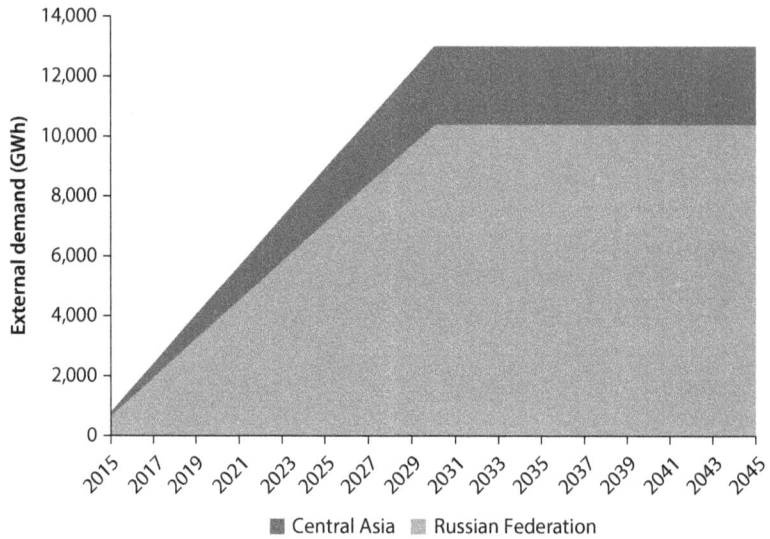

Note: GWh = gigawatt hours.

Figure 5.12 Annual External Capacity Required to Export Electricity to Central Asia and the Russian Federation, 2015–45

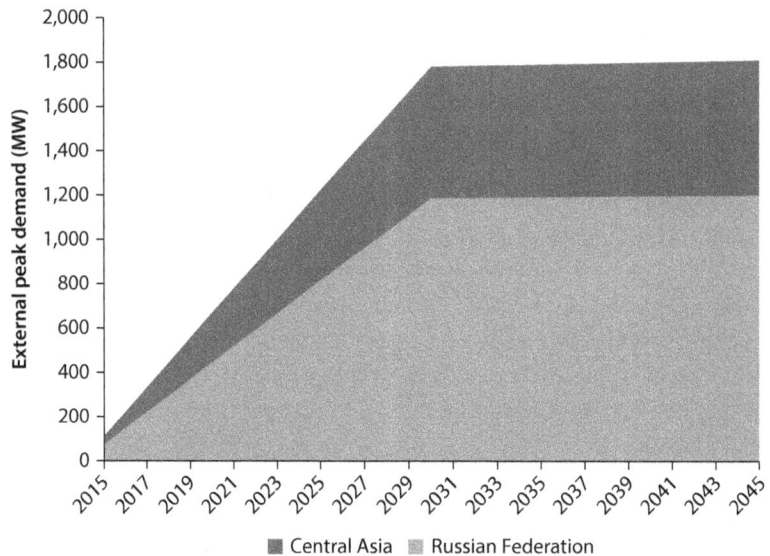

Note: MW = megawatts.

Least-Cost Case Scenario

(i)   *Internal interconnections.* The Least-Cost Case scenario is the only one that
       considers internal interconnections. Those transmission projects are not
       mandatory, but are subject to least-cost optimization. There is a minimum
       entry date constraint included; transmission projects can come no earlier
       than 2019, given the time needed for planning and construction. Because
       Kazakhstan's power system is modeled as a three-node system, the only
       transmission options are to (a) interconnect the Western and Northern
       zones; (b) interconnect the Western and Southern zones; and (c) strengthen
       the capacity of the North–South corridor.[12] The addition of internal trans-
       missions is not the only distinguishing difference of the Least-Cost Case
       scenario. Other than the Balkhash and Ekibastuz GRES extension, no other
       mandated projects or policy-related guidelines or targets are translated
       to mathematical constraints in the optimization process. Such constraints
       narrow the feasibility region and the chance to achieve a truly least-cost
       solution. The Least-Cost Case scenario is the least expensive of all the
       scenarios.

(ii)   *Sensitivity to natural gas price.* A variant of the Least-Cost Case scenario uses
       the economic cost of natural gas rather than the actual cost to identify the
       impact of natural gas pricing. The World Bank data suggest that KazTransGas
       buys from Russian Gazprom at US$85 per 1,000 m^3 of natural gas in the
       Southern zone, and returns the same amount of gas to Russia from the
       oil mines located in the Western zone (Karachaganak gas repository) at
       the same price. In this variation of the Least-Cost Case scenario, a homoge-
       neous border, price-based economic cost of natural gas is assumed at US$85
       per 1,000 m^3 (or around US$2.54 per GJ at the time this report was writ-
       ten). Since the cost of natural gas in the Southern zone is already US$2.54
       and no natural gas is available in the Northern zone, only the Western zone
       is affected by the change in the cost of natural gas.

Results and Discussion

Base Case Scenario

(i)   *Generation.* In the Base Case scenario, Kazakhstan's future power system is
       expected to be primarily coal-based. Coal-fired electricity generation will
       grow from about 70 TWh in 2015 to 100 TWh in 2045 (figure 5.13).
       However, the share of coal in the energy mix will drop over time (figure 5.14),
       because the growth in demand is higher than the growth in production
       from coal-fired units. Coal-fired units produce about 75 percent of total
       energy in 2015, but produce only 50 percent in 2045. Similarly, gas-fired
       generation increases from about 15 TWh in 2015 to 40 TWh in 2045
       (figure 5.13).

         The share of renewables in the energy mix grows over time. The
       share of PV and wind is 3 percent in 2010, 10 percent in 2030, and
       21 percent in 2045, as per the government's targets. The total share of

Figure 5.13 Generation, by Fuel Type (Base Case Scenario), 2015–45

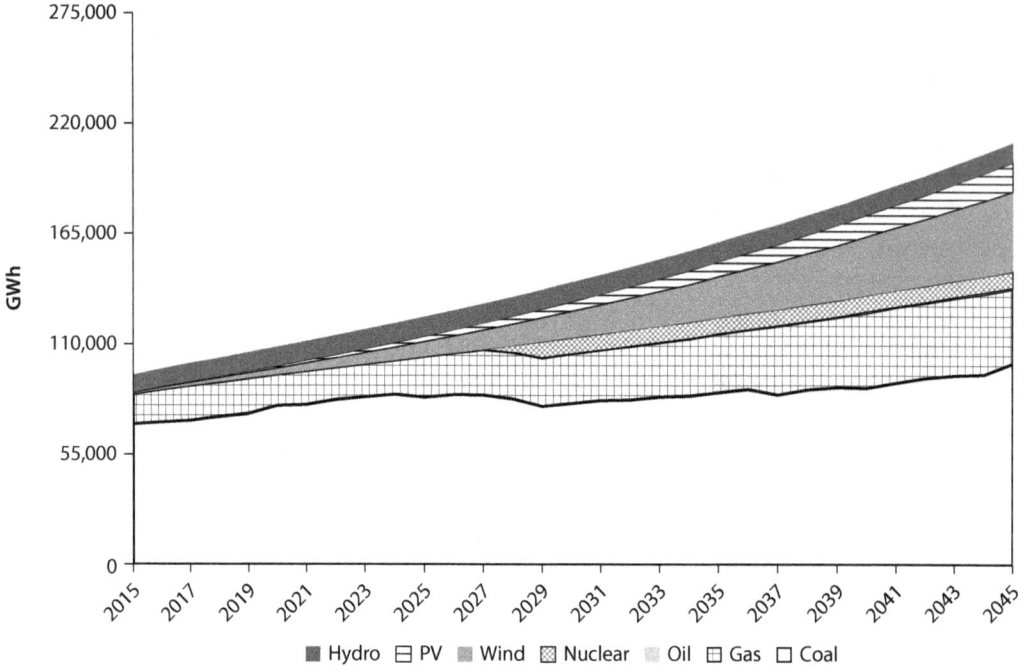

Note: GWh = gigawatt hours; PV = photovoltaic.

Figure 5.14 Thermal Generation, by Fuel Type (Base Case Scenario), 2015–45

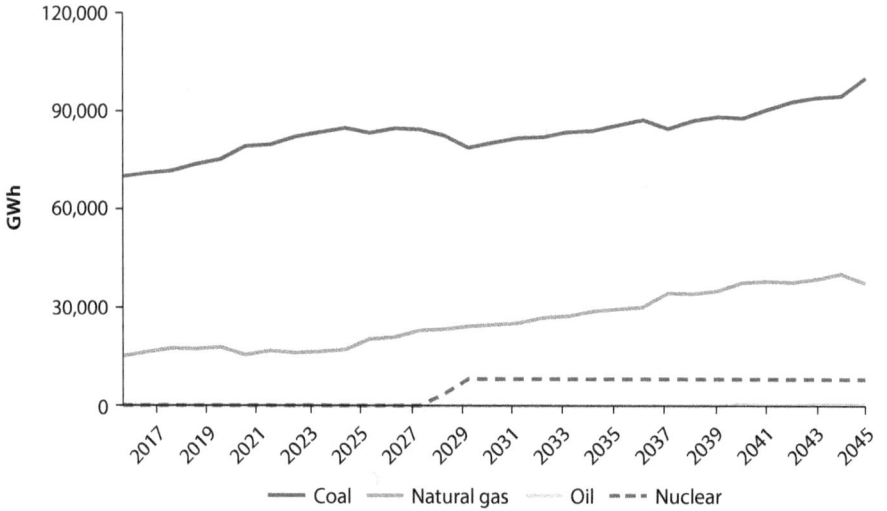

Note: GWh = gigawatt hours.

renewables (hydro included) is 20 percent in 2030 and 30 percent in 2045 (figures 5.15 and 5.16).

The share of electricity production from CHPs drops over time, although not significantly. CHPs produce 34 percent of total energy in 2015 and 29 percent in 2040. This occurs because the assumed growth in heating requirements is lower than the assumed growth in demand. In terms of nuclear power, 1 GW of nuclear power produces approximately 8 TWh of electricity per year.

(ii) *Installed capacity and reserve margin.* Most investments take place in the Northern zone, as expected. Almost all installed capacity in the Northern zone is coal-fired. The majority of investments in the Southern and Western zones are in gas-fired technologies (figure 5.17).

Approximately 3,500 MW of CHP coal-fired capacity and 500 MW of supercritical coal capacity will be needed in the Northern zone by 2030. In the Southern zone, the Balkhash supercritical coal-fired unit (1,320 MW) is almost the only coal-fired investment project. However, the Southern zone will need about 2 GW of gas-fired investments (CHP and OCGT) by 2030. Low gas prices in the Western zone favor CCGT investments for base load generation. By 2030, 1,800 MW of CCGT capacity will be needed, and by 2045, 3,200 MW of CCGT capacity will be needed.

Figure 5.15 Generation, by Fuel Type (Base Case Scenario), 2015–45

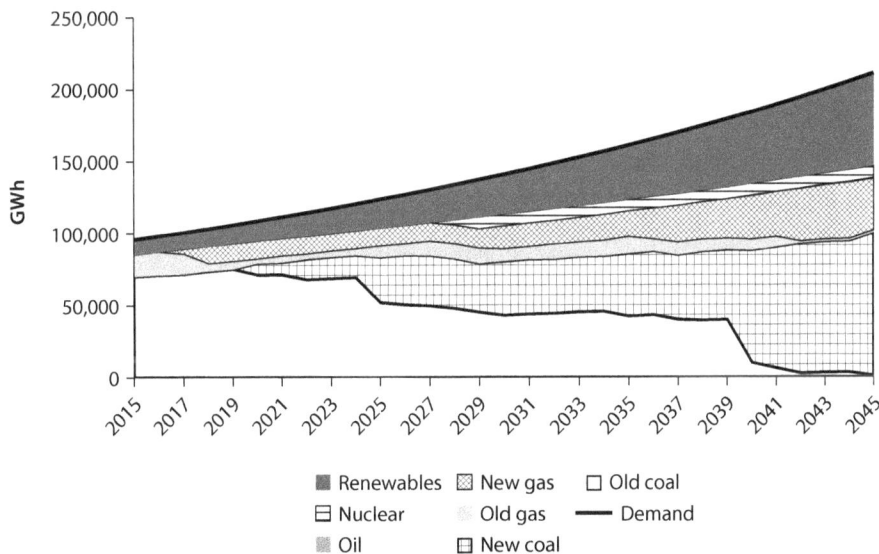

Note: GWh = gigawatt hours. Coal- and gas-fired generation is broken down into "old" and "new." Old coal represents production from existing coal-fired generators, while new coal represents generation from future new and more efficient units yet to come on line. The same applies for gas-based generators.

Figure 5.16 Generation, by Fuel as a Percentage of Total Generation (Base Case Scenario), 2015–45

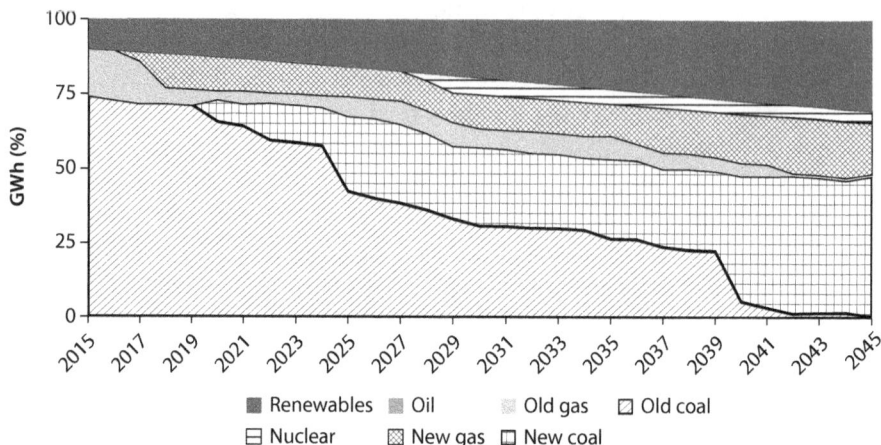

Note: GWh = gigawatt hours.

Figure 5.17 Installed Thermal Capacity, by Technology and Electrical Zone (Base Case Scenario), 2019–49

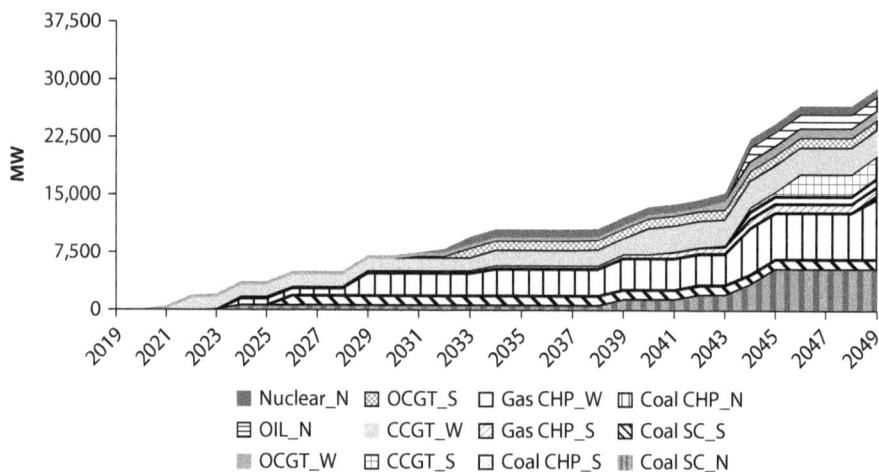

Note: CCGT = combined cycle gas turbine; CHP = combined heat and power plant; MW = megawatts; N = Northern zone; OCGT = open cycle gas turbine; S = Southern zone; SC = supercritical; W = Western zone.

The rehabilitation/extension program, together with new greenfield investments, will contribute to bringing the reserve margin to healthy levels. The reserve margin will grow to 18 percent in 2018, and will remain above that level for the rest of the optimization period (figure 5.18).

(iii) *Capital investments.* The undiscounted total capital cost of the expansion program will be US$42 billion by 2030 and US$99 billion by 2045 (figures 5.19 and 5.20). Rehabilitation/extensions are the main forms of

Figure 5.18 Available Capacity, by Fuel Type and Reserve Margin (Base Case Scenario), 2015–45

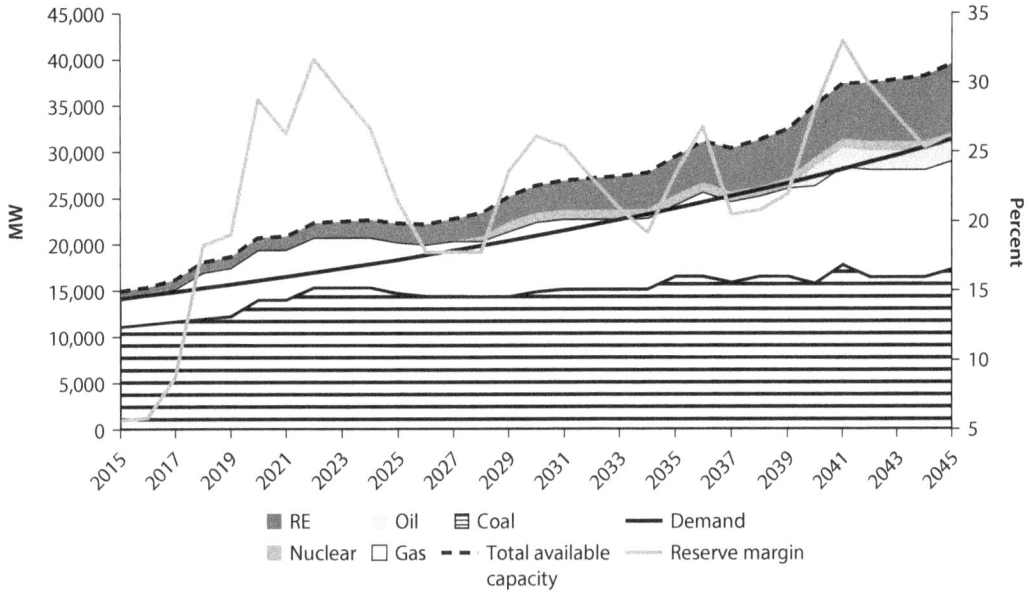

Note: MW = megawatts; RE = renewable energy.

Figure 5.19 Cumulative Undiscounted Capital Costs for the Generation Expansion Program (Base Case scenario), 2015–45

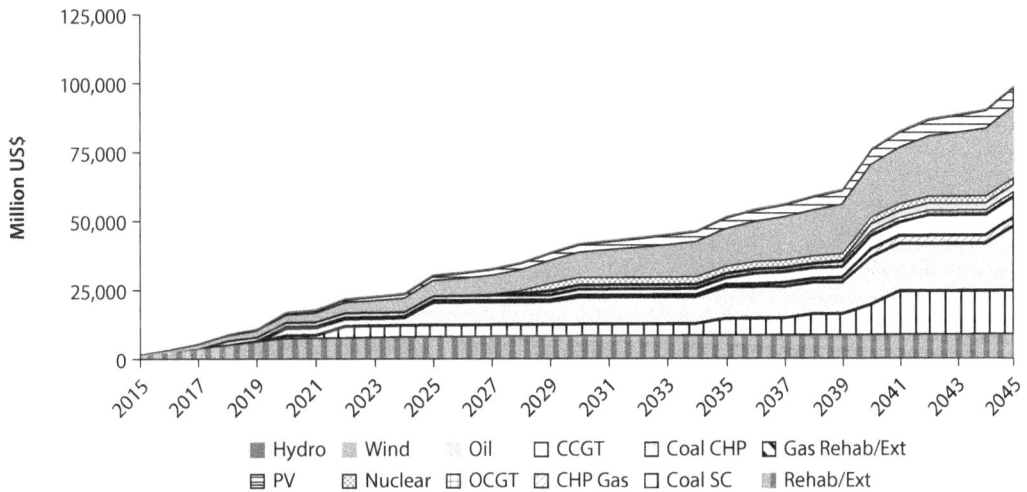

Note: CCGT = combine cycle gas turbine; CHP = combined heat and power plant; OCGT = open cycle gas turbine; PV = photovoltaic; SC = supercritical.

Figure 5.20 Estimated Investment Requirements, 2015–30

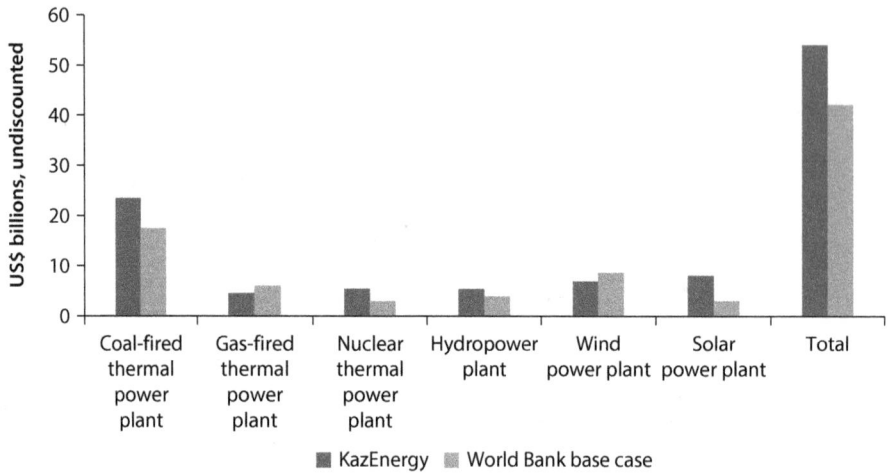

investments up to 2020 (US$7.2 billion). Investments in coal-fired tech-
nologies account for 50 percent of the total cost of capital investments
in 2030.

The Southern zone, consuming only 20 percent of total demand,
requires 33 percent of total investments in new generation projects
(table 5.6), because most installations of expensive PV and wind projects take place there. More than 50 percent of total investments take
place in the Northern zone, as expected. As shown in table 5.7, about
60 percent of total investments take place during the second half of the
planning period (2031–45). The total annualized undiscounted investment cost for the Base Case scenario is US$81.5 billion. The Kazakhstan
power system's LCOE is US$31.5/kilowatt hour.

(iv)   *Operational costs.*[13] The operational costs of the system grow with time
       because of demand growth—from US$21 billion in 2015 to US$38 billion
       in 2045 (figure 5.21). The total cumulative operational costs are US$36
       billion by 2030 and US$85 billion by 2045.

       However, the annual average systemwide operational cost in US$/
       megawatt hour (MWh) drops over time as new and more efficient units
       come on line to replace old inefficient ones. The average systemwide cost
       over the planning period (2015–45) is US$19.2/MWh.

(v)   *Systemwide LCOE.* Kazakhstan's power system's LCOE is US$35.1/MWh
       based on annualized capital investments, operational costs, and energy
       produced—all discounted at 6 percent.

(vi)   *Emissions.* CO_2 emissions remain constant at about 90 million tons per year
       over the planning period, although demand grows over time (figure 5.22).

Table 5.6 Overnight Undiscounted Capital Investments in Generation Projects, by Zone (Base Case Scenario)
(US$ millions)

Investment	North	South	West
Combined heat and power: coal	22,184	1,165	0
Supercritical	12,758	3,168	0
Combined heat and power plant: gas	0	1,637	1,558
Combined cycle gas turbine	0	3,588	3,924
Open cycle gas turbine	0	1,161	837
Oil	2,000	0	0
Nuclear	3,000	0	0
Wind	7,319	14,099	3,992
Photovoltaic	505	4,768	1,889
Hydro	518	202	0
Total	48,284	29,786	12,201

Note: The values are for the entire planning period (2015–45). "Overnight capital investments" (or "costs") is a term used in the power generation industry to describe the capital costs of construction, excluding the financing costs. This table does not include the cost of rehabilitation/extension of generating units.

Table 5.7 Overnight Undiscounted Capital Investments in New Generation and Rehabilitation (Base Case Scenario)
(US$ millions)

Investment	2015–20	2021–25	2026–30	2031–45
Combined heat and power plant: coal	2,694	5,538	1,529	13,588
Supercritical	1,260	3,168	0	11,498
Combined heat and power plant: gas	421	141	153	2,479
Open cycle gas turbine	332	0	1,161	506
Combined cycle gas turbine	1,734	0	457	5,321
Oil	0	0	0	2,000
Nuclear	0	0	3,000	0
New wind	2,116	2,878	3,633	16,782
New photovoltaic	904	1,105	1,056	4,096
New hydro	720	0	0	0
Rehabilitation/extension of generation assets	7,266	692	176	528
Zonal interconnections	0	0	0	0
Total	17,447	13,523	11,165	56,798

Stabilization of emissions is a consequence of growth of renewable energy over time, and the decommissioning of old, inefficient units and replacement with new, efficient ones.

The systemwide emissions intensity drops drastically over time—from 0.98 ton CO_2 per MWh in 2015 to 0.45 ton CO_2 per MWh in 2045.

Figure 5.21 Operational Costs, by Fuel (Base Case Scenario), 2015–45

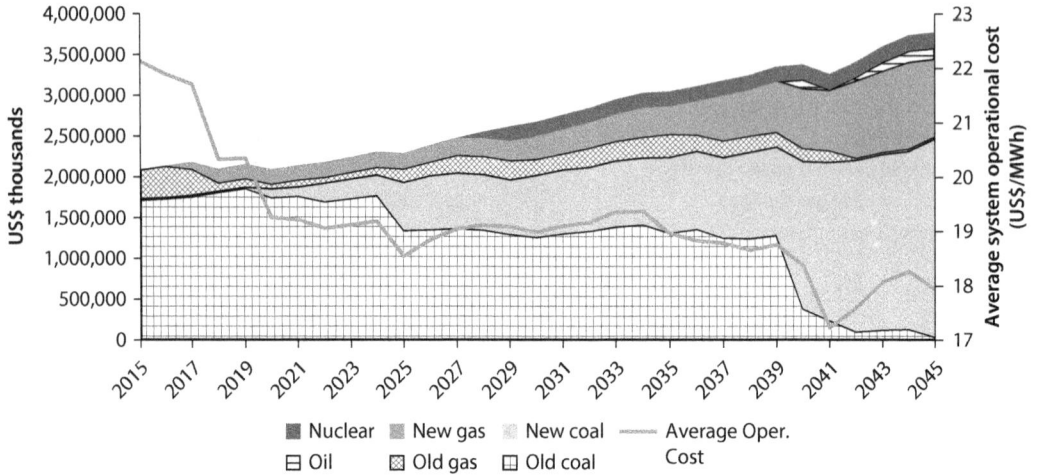

Figure 5.22 Carbon Dioxide Emissions and Systemwide Emissions Intensity (Base Case Scenario), 2015–45

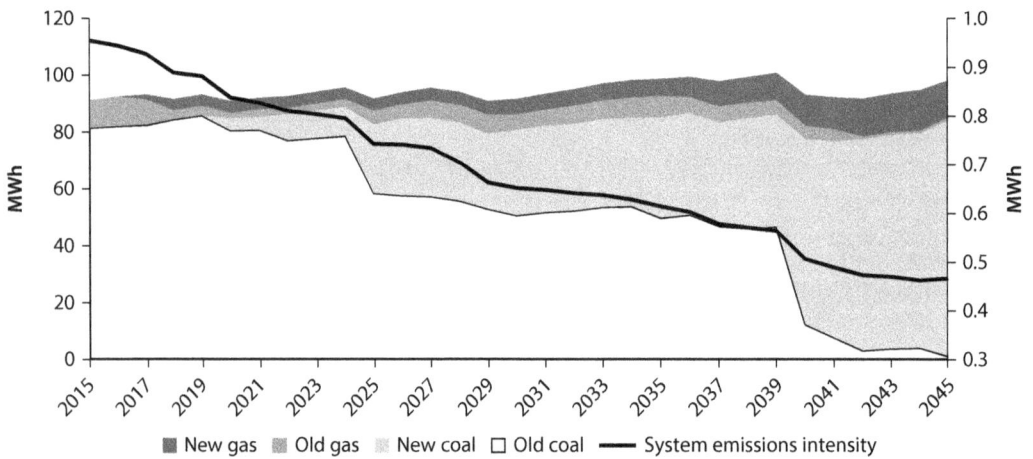

Note: CO_2 = carbon dioxide; MWh = megawatt hours.

Green Case Scenario

(i)   *Generation.* In the Green Case scenario, total coal-fired generation decreases over time (figures 5.24 and 5.25). Coal production, on the one hand, drops to 50 TWh at the end of the planning study from 70 TWh in the beginning (figures 5.23 through 5.25).

On the other hand, generation from gas-fired units increases from about 15 TWh in 2015 to 34 TWh in 2045. Gas-fired generation doubles in 2020 when the conversion of coal-fired to gas-fired CHPs takes place in the Northern zone.

Figure 5.23 Generation, by Fuel Type (Green Case Scenario), 2019–49

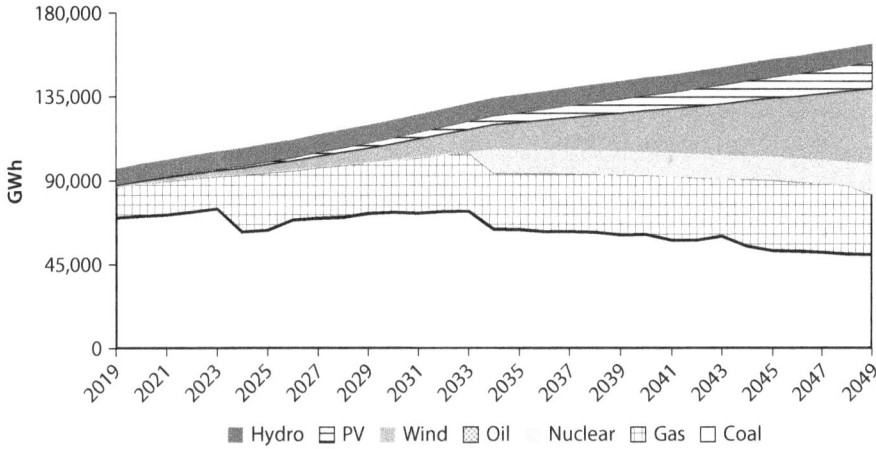

Note: GWh = gigawatt hours; PV = photovoltaic.

Figure 5.24 Thermal Generation, by Fuel Type (Green Case Scenario), 2015–45

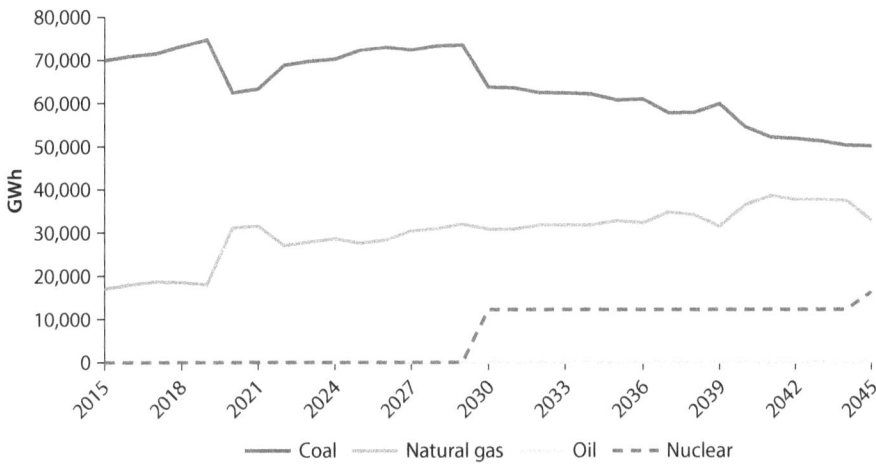

Note: GWh = gigawatt hours.

At the end of the planning period, about 30 percent of total generation comes from coal-fired units and 20 percent from gas-fired units, while the rest comes from carbon-free technologies (renewables and nuclear).

(ii) *Installed capacity.* The expansion plan in the Green Case scenario requires about 3 GW less thermal installed capacity than in the Base Case scenario (28 GW versus 31 GW). About 50 percent of total installed capacity at the end of the planning period will be coal-fired generation. Gas-fired generation accounts for 42 percent and nuclear 8 percent of new thermal installed capacity. About 3.3 GW, or 15 percent of total installed capacity, will be OCGT. OCGT capacity is required to balance variable renewable energy generation. The effect of solar and wind variability is higher in the

Figure 5.25 Generation, by Fuel Type as a Percentage of Total Generation (Green Case Scenario), 2015–45

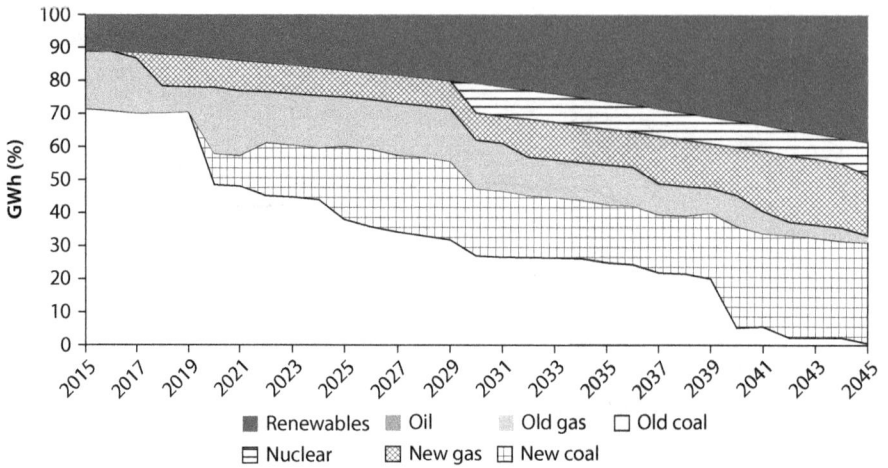

Renewables   Oil   Old gas   □ Old coal
⊟ Nuclear   New gas   ⊞ New coal

Note: GWh = gigawatt hours.

Figure 5.26 Cumulative Undiscounted Capital Costs for the Generation Expansion Program (Base Case Scenario), 2015–45

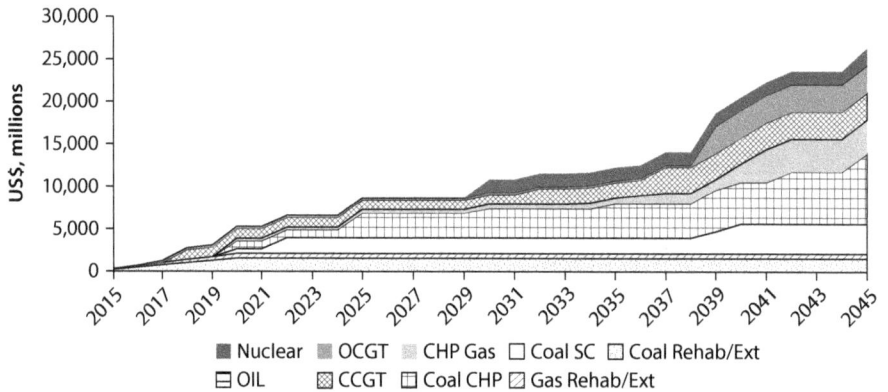

■ Nuclear   ▨ OCGT   CHP Gas   □ Coal SC   □ Coal Rehab/Ext
⊟ OIL   ▨ CCGT   ⊞ Coal CHP   ▨ Gas Rehab/Ext

Note: CCGT = combined cycle gas turbine; CHP = combined heat and power plant; MW = megawatts; OCGT = open cycle gas turbine; SC = supercritical.

Green Case scenario compared with the Base Case scenario, because of the higher penetration of renewables in the energy mix.

(iii) *Capital investments.* Total undiscounted capital investments for generation expansion in the Green Case scenario are US$91.5 billion—US$7.4 billion less than in the Base Case scenario (figures 5.26 and 5.27). About 50 percent of investments will be for coal-fired technologies. The total annualized capital investment during the planning period will be US$78.4 billion—US$3.1 billion less than in the Base Case scenario (table 5.8).

Figure 5.27 Cumulative Undiscounted Capital Costs for the Generation Expansion Program (Green Case Scenario), 2015–45

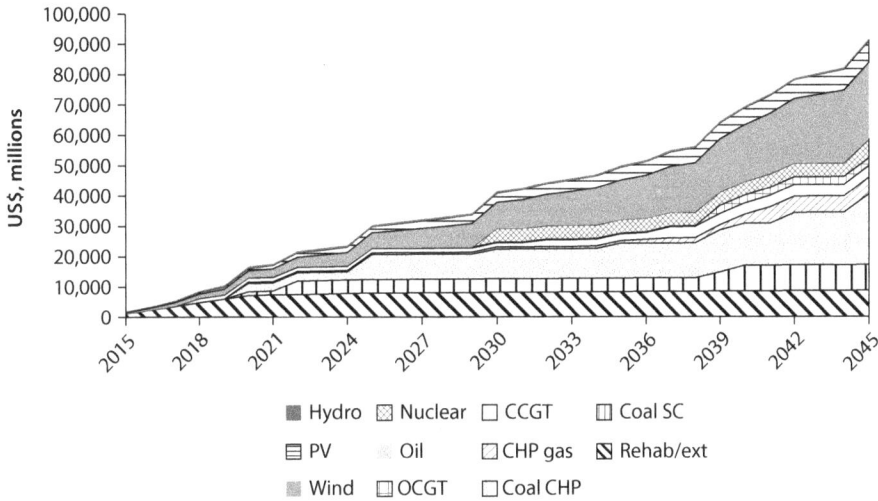

Note: CCGT = combined cycle gas turbine; CHP = combined heat and power plant; OCGT = open cycle gas turbine; PV = photovoltaic; SC = supercritical.

Table 5.8 Difference in Capital Investments between Green and Base Case Scenarios
(US$ millions)

	2015–20	2021–25	2026–30	2031–45
Combined heat and power plant: coal	0	0	0	−61
Supercritical	0	0	0	−7,509
Combined heat and power plant: gas	0	0	0	2,339
Open cycle gas turbine	−151	0	−1,161	1,894
Combined cycle gas turbine	−356	0	−457	−2,893
Oil	0	0	0	−2,000
Nuclear	0	0	1,500	1,500
New wind	0	0	0	0
New photovoltaic	0	0	0	0
New hydro	0	0	0	0
Rehabilitation/extension of generation assets	0	0	0	0
Zonal interconnections	0	0	0	0
Total	**−507**	**0**	**−118**	**−6,730**

Note: Negative values indicate savings achieved in the Green Case scenario.

(iv) *Operational costs.* Total operational costs over the planning period are US$82.4 billion—US$2.9 billion less than in the Base Case scenario (figure 5.28).

The average operational costs of the Green Case scenario are US$20.4/ MWh—US$1.2/MWh more than in the Base Case scenario. The result seems counterintuitive considering the reduction in demand caused by

Figure 5.28 Operational Costs, by Fuel (Green Case Scenario), 2015–45

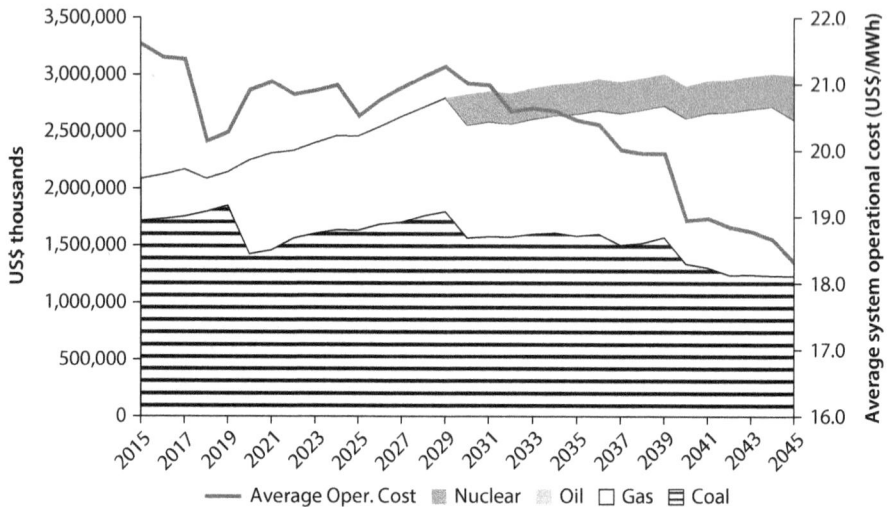

Note: MWh = megawatt hours.

energy efficiency in the Green Case scenario. The difference is related to gasification of the Northern zone. About 1.8 GW of coal-fired capacity is converted to CHP gas-fired capacity. The cost of gas in the Northern zone is almost four times as much as the cost of displaced coal. When dividing total cost by energy produced, the average cost is more expensive compared with that seen in the Base Case scenario.

(v)   *Systemwide LCOE.* The systemwide LCOE for the Green Case scenario is US$38.2/MWh, based on annualized capital investments, operational costs, and energy produced—all discounted at 6 percent. However, this cost does not account for the cost of energy efficiency measures, because annual investment figures for the energy efficiency program were not available.

(vi) *Emissions.* Emissions are drastically reduced in the Green Case scenario (figure 5.29). Total CO_2 emissions savings over the planning period, compared with the Base Case scenario, equal half a billion tons. The system's emissions drop from 90 million tons in 2015 to 54.5 million tons in 2045, while the system's emissions intensity drops from 0.95 to 0.34 tons of CO_2 per MWh. This is a 66 percent reduction in emissions intensity, attributable to the growing penetration of renewable energy in the system. The emissions savings in 2045 are 40 percent less than those in 2012 (figure 5.30).

(vii) *Comparison of the Green and Base Case scenarios.* The Green Case scenario has higher LCOE than the Base Case scenario (US$38.2/MWh compared with US$35.1/MWh), although the total costs (total cumulative annualized capital investments plus total operational costs over the planning

Figure 5.29 Intensity of Carbon Dioxide Emissions and Systemwide Emissions (Green Case Scenario), 2015–45

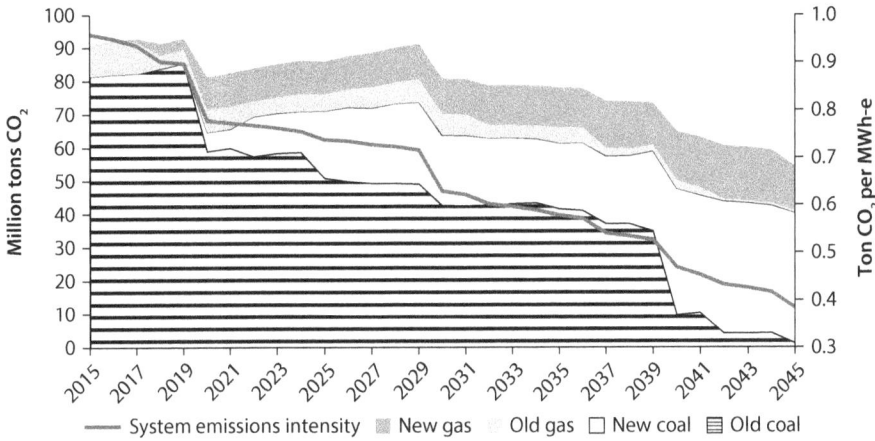

Note: CO_2 = carbon dioxide; MWh-e = megawatt hours of electricity.

Figure 5.30 Green Case Scenario: Reduction in Carbon Dioxide Emissions Compared with 2012 Levels, 2015–45

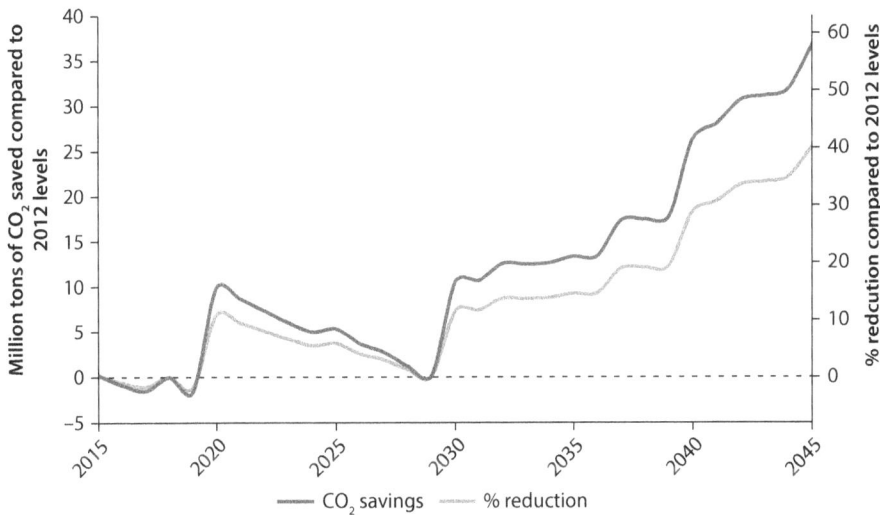

Note: CO_2 = carbon dioxide.

period for the Green Case scenario) are US$6 billion less compared with the Base Case scenario. Two factors contribute to the Green Case scenario's higher LCOE but lower system costs: a high penetration of variable renewable energy sources, which requires balancing thermal units to run at low capacity factors, and high operational costs because of the gasification of the Northern zone. The former requires higher firm capacity to produce the same amount of energy, while the latter imposes suboptimal dispatch.

The LCOE calculation for the Green Case scenario does not account for the cost of investment in energy efficiency that led to a reduction in demand. Factoring this in, the LCOE calculation requires annual investment figures on energy efficiency—data that are lacking. However, knowing that the Green Case scenario is US$6 billion less expensive than the Base Case scenario, and the demand reduction over the planning period is 458 TWh, the cost of the energy efficiency program must be US$13/MWh or less for the Green Case scenario to become less expensive than the Base Case scenario. In absolute terms, the LCOE will still be higher. Most energy efficiency programs cost about US$50 to US$100/MWh (Spees and Lave 2007)[14] and, thus, it is very unlikely that the Green Case scenario will cost less than the Base Case scenario.

However, the Green Case scenario has significant environmental benefits because of the CO_2 savings and reduced air pollution in cities, resulting from conversion of CHPs from coal-fired to gas-fired.

Global CO_2-related externalities are embedded in the suggested carbon prices by the World Bank (World Bank 2014) (table 5.9). If carbon pricing is factored into the LCOE calculation for the Green Case scenario, the LCOE value is US$29.7/MWh. The total undiscounted savings from CO_2 reductions equal US$28 billion over the planning period. Accounting for externalities, any energy efficiency program that costs US$75/MWh or less, on average, would render the cost of the Green Case scenario lower than the cost of the Base Case scenario in absolute terms (it is possible that the LCOE still will be higher when the cost of energy efficiency is factored in).

It was not possible to quantify the environmental benefits caused by local externalities avoided from conversion of CHPs from coal-fired to gas-fired. This was because of the lack of relevant data.

The analysis reveals that the Green Case scenario is not economical in the absence of carbon pricing. To account for global CO_2-related externalities, the World Bank's social value of carbon was used to calculate the system-wide LCOE after the capacity-planning optimization routine was complete.

Table 5.9 Carbon Prices Based on World Bank Guidelines
real 2014 U.S dollar/metric ton of carbon dioxide equivalent

	2015	2020	2030	2040	2050
Low	15	20	30	40	50
Base	30	35	50	65	80
High	50	60	90	120	150

Source: World Bank.

This pricing scheme may also be incorporated into the planning exercise by treating it as an additional variable cost to generation. It is expected that such an analysis would result in the increased development of "greener" technologies, but it is not clear by how much. Therefore, in the future it may be useful to incorporate this carbon pricing scheme into the objective function of the optimization routine and compare the results of the proposed analysis with those of the pure least-cost scenario, or observe the carbon price at which it becomes economical to meet the renewable targets of the Green Case scenario.

Regional Export Case Scenario

(i) *Generation.* Similar to the Base Case scenario, coal-based generation is dominant, followed by natural gas. Generation from both sources increases over time, although their penetration drops as demand grows. The main difference between the two scenarios is that external demand has to be satisfied through additional generation (figure 5.31). Figure 5.31 reflects that the additional generation comes mainly from coal-fired power plants.

(ii) *Installed capacity.* The expansion plan in the Regional Export Case scenario requires 1.5 GW of thermal installed capacity in addition to the capacity installed in the Base Case scenario (30.5 GW versus 29 GW) (figures 5.33 and 5.34). One of the observations from figure 5.32 is the need for supercritical coal-fired capacity in the Northern zone to produce the bulk of external demand in a least-cost manner. A second implication is the increased need for flexible generation in the Southern zone. The Regional Export Case scenario requires an additional 1,800 MW of OCGT capacity in the Southern zone compared with the Base Case scenario. This is because the Southern zone allocates some of its base capacity to satisfy an external

Figure 5.31 Generation, by Fuel (Regional Export Case Scenario), 2015–45

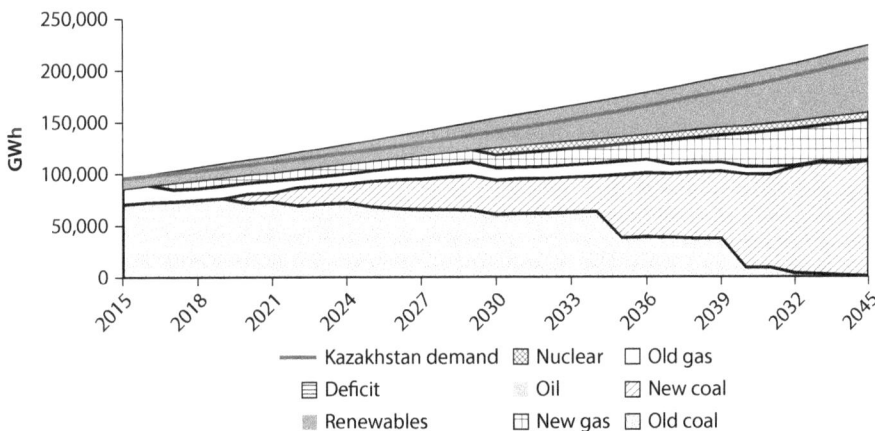

Note: GWh = gigawatt hours.

Figure 5.32 Comparison of Generation from Coal-Fired and Gas-Fired Power Plants between the Regional Export and Base Case Scenarios, 2015–45

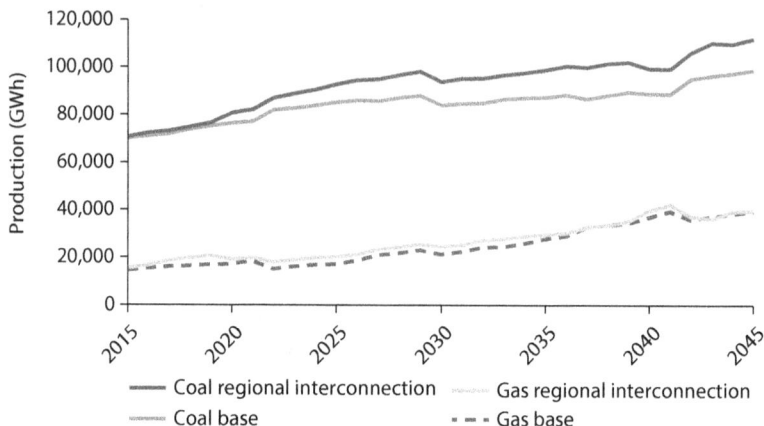

Note: GWh = gigawatt hours.

Figure 5.33 Thermal-Installed Capacity, by Technology (Regional Export Case Scenario), 2015–45

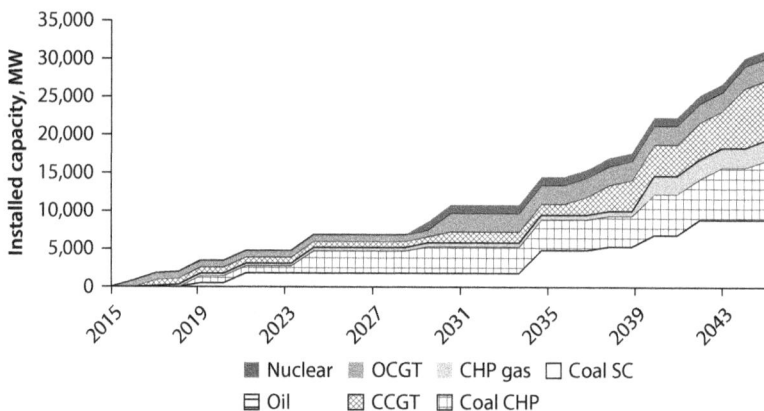

Note: CCGT = combined cycle gas turbine; CHP = combined heat and power plant; MW = megawatts; OCGT = open cycle gas turbine; SC = supercritical.

flat load for six months a year; at the same time, the Southern zone receives most of the variable renewable energy generation. OCGT is used to provide the required system flexibility rather than as a least-cost generation solution.

(iii) *Capital investments.* Total undiscounted capital investments for generation expansion in the Regional Export Case scenario equal US$102.3 billion—US$3.4 billion more than in the Base Case scenario (table 5.10).

Figure 5.34 Difference in Installed Capacity between the Regional Export and Base Case Scenarios

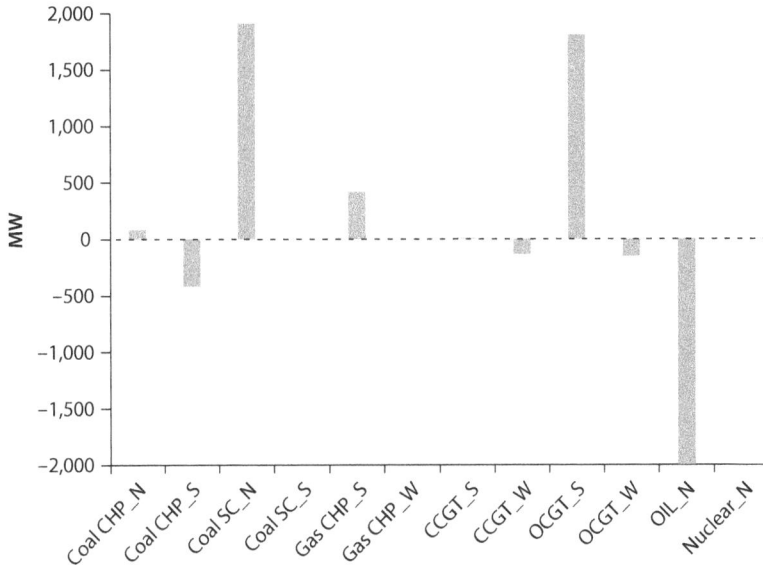

Note: Positive values indicate additional capacity in the Regional Export Case scenario. CCGT = combined cycle gas turbine; CHP = combined heat and power plant; MW = megawatts; OCGT = open cycle gas turbine; N = North; S = South; SC = supercritical; W = West.

Table 5.10 Difference in Undiscounted Capital Investments in Generation Projects between the Base and Regional Export Case Scenarios
(US$ millions)

	2015–20	2021–25	2026–30	2031–45
Combined heat and power plants: coal	0	0	0	−935
Supercritical	0	0	0	4,584
Combined heat and power plants: gas	0	0	0	582
Open cycle gas turbine	383	0	−206	1,147
Combined cycle gas turbine	−978	0	−457	1,284
Oil	0	0	0	−2,000
Nuclear	0	0	0	0
New wind .	0	0	0	0
New photovoltaic	0	0	0	0
New hydro	0	0	0	0
Rehabilitation/extension of generation assets	0	0	0	0
Zonal Interconnections[a]	0	0	0	0
Total	**−596**	**0**	**−664**	**4,662**

a. Kazakhstan's power system was simulated as a three-node system. The model cannot capture the need to strengthen transmission capacity within each zone, so there is reduced insight on transmission investments. The model only captures the need for transmission capacity between zones.

The total annualized investment over the planning period is US$83.3 billion—US$1.6 billion more than in the Base Case scenario. Most additional investments, compared with the Base Case scenario, are required after 2030.

(iv) *Operational costs.*[13] Total operational costs over the planning period are US$93.6 billion—US$8.3 billion less than in the Base Case scenario. The average operational system cost is US$19.5/MWh—slightly higher than in the Base Case scenario, because renewables in this scenario have a lower share in total production.

(v) *Systemwide LCOE.* The systemwide LCOE for the Regional Export Case scenario is US$34.8/MWh—slightly lower compared with the Base Case scenario's LCOE.

(vi) *Emissions.* The Regional Export Case scenario has a slightly higher emissions intensity compared with that of the Base Case scenario (0.71 versus 0.69 ton CO_2/MWh-e) since additional external demand is supplied by emissions-intensive, coal-fired units. In total, CO_2 emissions in the Regional Export scenario are about 300 million tons more than in the Base Case scenario.

(vii) *Comparison with Base Case scenario.* The Regional Export scenario has lower LCOE than in the Base Case scenario, although it costs US$9.9 billion more (undiscounted annualized investment costs + undiscounted operational costs). The lower LCOE is related to additional external demand being supplied by coal-fired units having the lowest technology-related (not systemwide) LCOE in the Northern zone at high capacity factors.[15] However, investments in additional capacity in the Regional Export Case scenario are made to produce electricity that can be traded, thereby earning profits for the system. The break-even cost of exported electricity is US$33/MWh. This means that electricity should be exported at a price higher than 3.3 U.S. cents/kWh to make a profit.

Least-Cost Case Scenario

(i) *Transmission.* One of the main differences between the Least-Cost Case scenario and the others is that it optimizes internal transmission. More specifically, the model explored whether it is economical to interconnect the Western zone with the Northern and Southern zones (figure 5.35). The cost of interconnection was assumed to be US$138/kW and US$156/kW for the North–West and South–West interconnections, respectively.[16] Total project capacity was limited to 10 GW, and the final capacity was subject to economic optimization.

The existing current capacity of the North–South transmission corridor is 1,350 MW and is expected to increase to 2,100 MW by 2018. The possibility of strengthening the current interconnection capacity, at a cost of US$100/kW, in the future was considered. Total final capacity is limited to 10 GW.

Figure 5.35 Interzonal Transmission Capacity Requirements, 2015–45

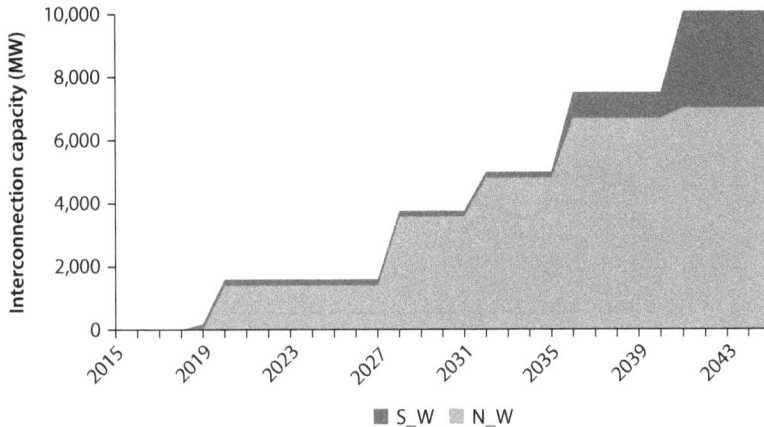

Note: MW = megawatts; N_W = North–West; S_W = South–West.

The results show that current plans for the North–South interconnection (2.1 GW by 2018) are sufficient over the entire optimization period; no additional capacity is required up to 2045.

As discussed in the next section, however, significant amounts of gas-related power flows from the Western zone toward the Northern and Southern zones. Up to about 7.0 GW of North–West interconnection capacity will be needed by 2041 to transfer gas-related power to the Northern zone. In addition, South–West transmission capacity of up to 2 GW will be required near the end of the optimization period for similar reasons.

(ii) *Generation.* In the Least-Cost Case scenario, generation from natural gas-fired units becomes dominant over time (from 16 percent penetration in 2015 to 72 percent in 2045) (figure 5.36). Conversely, coal-fired penetration drops significantly over time—from 75 percent in 2015 to 30 percent by 2045 (figure 5.37). At the same time, wind never reaches grid parity. Moreover, solar power becomes economical in 2041 and supplies up to 13 percent of total energy by 2045. Comparison of Generation from Coal- and Gas-Fired Sources between the Base and Least-Cost Case Scenarios is reflected in figure 5.38.

(iii) *Installed capacity.* Growth of gas-fired generation is based on the low fuel prices in the western part of the country where the oil fields are located. The least-cost analysis shows that it is worth investing in long transmission lines to transfer gas-fired electricity from the Western zone to the rest of the country. As shown in figure 5.39, the Least-Cost Case scenario requires about 30 GW of new thermal capacity and another 10 GW of PV by the end of the planning period. About 37 percent of total new thermal capacity is CCGT, placed in the Western zone.

Figure 5.36 Generation, by Fuel Type (Least-Cost Case Scenario), 2015–45

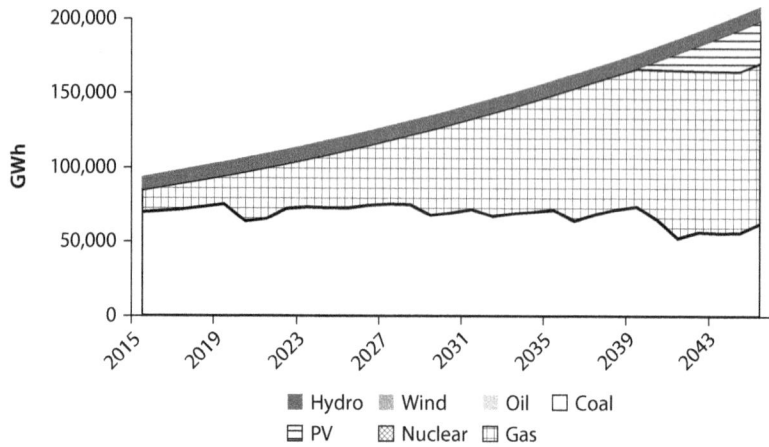

Note: GWh = gigawatt hours; PV = photovoltaic.

Figure 5.37 Comparison of Generation from Coal- and Gas-Fired Sources between the Base and Least-Cost Case Scenarios, 2015–45

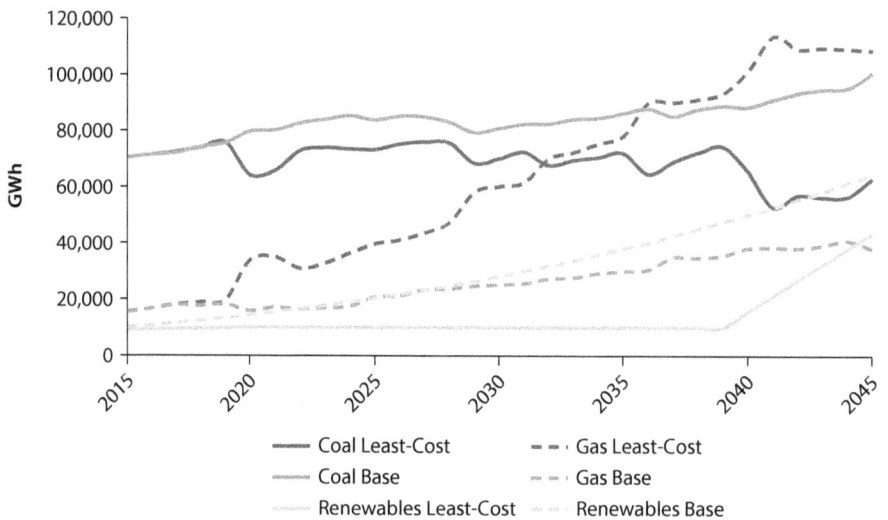

(iv) *Capital investments.* Total undiscounted capital investments for generation and transmission expansion in the Least-Cost Case scenario equal US$68.6 billion—US$30 billion less than in the Base Case scenario (table 5.11). The total annualized investment over the planning period is US$54.6 billion—US$27 billion less than in the Base Case scenario (table 5.12). Most investments are in coal-fired CHPs and CCGT technologies.

Figure 5.38 Generation, by Fuel Type as a Percentage of Total Generation (Least-Cost Case Scenario), 2015–45

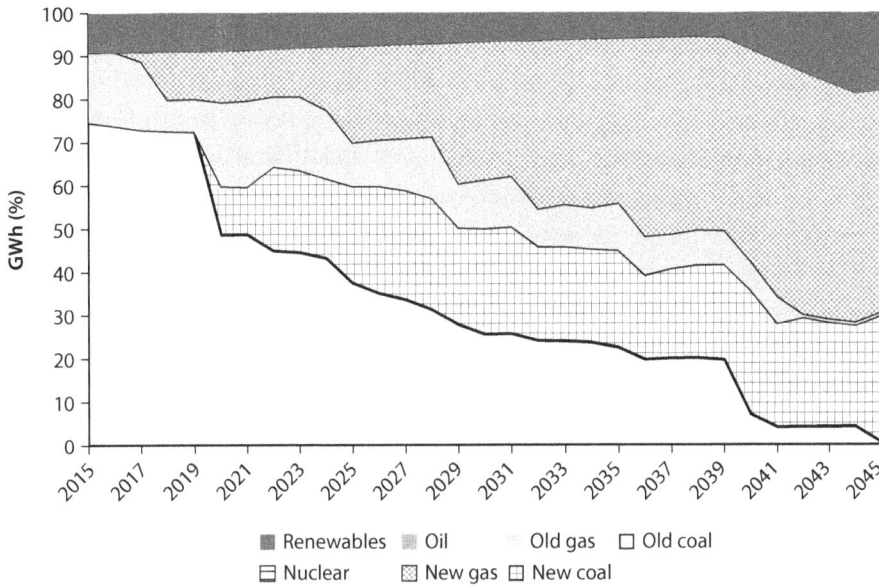

Figure 5.39 Installed Generation Capacity, by Technology (Least-Cost Case Scenario), 2015–45

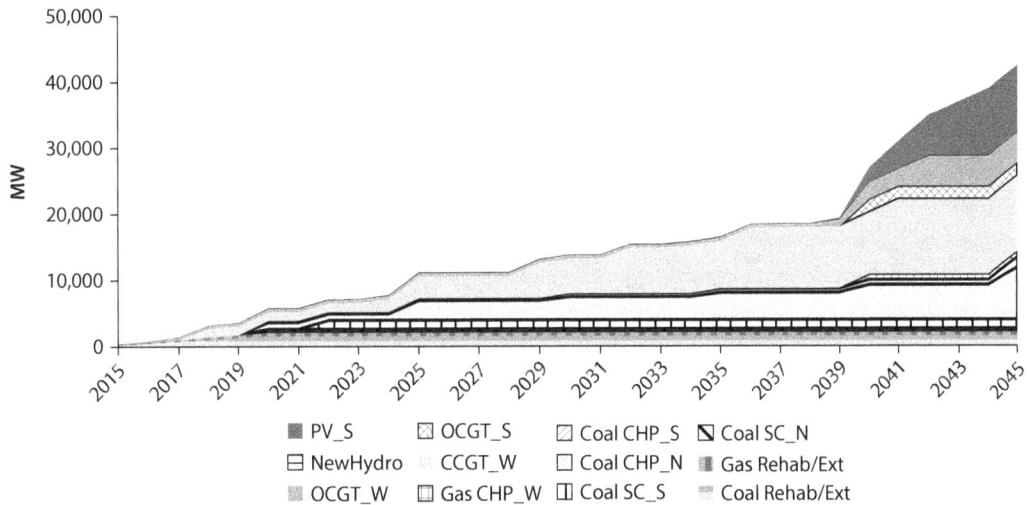

Note: CCGT = combined cycle gas turbine; CHP = combined heat and power plant; MW = megawatts; N = North; OCGT (open cycle gas turbine); PV = photovoltaic; S = South; SC = supercritical; W = West.

Table 5.11 Total Capital Investments in New Generation Technologies, Transmission, and Rehabilitation (Least-Cost Case Scenario)
(US$ millions)

	2015–20	2021–25	2026–30	2031–45
Combined heat and power plants: coal	2,694	5,538	1,529	12,244
Supercritical	1,260	3,168	0	0
Combined heat and power plants: gas	421	141	153	2,822
Open cycle gas turbine	221	0	0	4,842
Combined cycle gas turbine	1,620	2,361	2,400	7,200
Oil	0	0	0	0
Nuclear	0	0	0	0
New wind	0	0	0	0
New photovoltaic	0	0	0	9,185
New hydro	720	0	0	0
Rehabilitation/extension of generation assets	7,266	692	176	528
Zonal interconnections	223	0	301	922
Total	**14,424**	**11,900**	**4,558**	**37,744**

Table 5.12 Difference in Capital Investments in Generation Technologies between the Least-Cost and Base Case Scenarios
(US$, millions)

	2015–20	2021–25	2026–30	2031–45
Combined heat and power plants: coal	0	0	0	−1,344
Supercritical	0	0	0	−11,498
Combined heat and power plants: gas	0	0	0	343
Open cycle gas turbine	−111	0	−1,161	4,337
Combined cycle gas turbine	−114	2,361	1,943	1,879
Oil	0	0	0	−2,000
Nuclear	0	0	−3,000	0
New wind	−2,116	−2,878	−3,633	−16,782
New photovoltaic	−904	−1,105	−1,056	5,089
New hydro	0	0	0	0
Rehabilitation/extension of generation assets	0	0	0	0
Zonal interconnections	223	0	301	922
Total	**−3,023**	**−1,623**	**−6,607**	**−19,054**

Note: Negative values indicate savings for the Least-Cost Case scenario.

(v) *Operational costs.*[13] Total operational costs over the planning period are US$91.1 billion. This is US$5.8 billion less than in the Base Case scenario (figure 5.40).

    The average operational cost over the entire period is US$20.6/MWh versus US$19.2/MWh for the Base Case scenario. Higher operational costs are related to increased use of natural gas, which is more expensive than coal.

(vi) *Emissions.* One of the most important findings of the Least-Cost Case scenario analysis is that emissions are very similar to those in the Base Case scenario (about 2,900 million tons), although with much less investment in variable renewable energy technologies (figure 5.41).

(vii) *Systemwide LCOE.* The systemwide LCOE for the Least-Cost Case scenario is US$31.1/MWh. This is the lowest of all the scenarios, as expected.

(viii) *Comparison with the Base Case scenario.* CO_2 emissions in the Least-Cost Case scenario are very similar to those in the Base Case scenario. This means that investing in a total of 10 GW of high-voltage transmission capacity to interconnect the three zones for a total cost of about US$1.5 billion and only 10 GW of solar later in the planning period (US$9 billion worth) leads to CO_2 savings

Figure 5.40 Dynamics of Cumulative Operational Costs, 2015–45

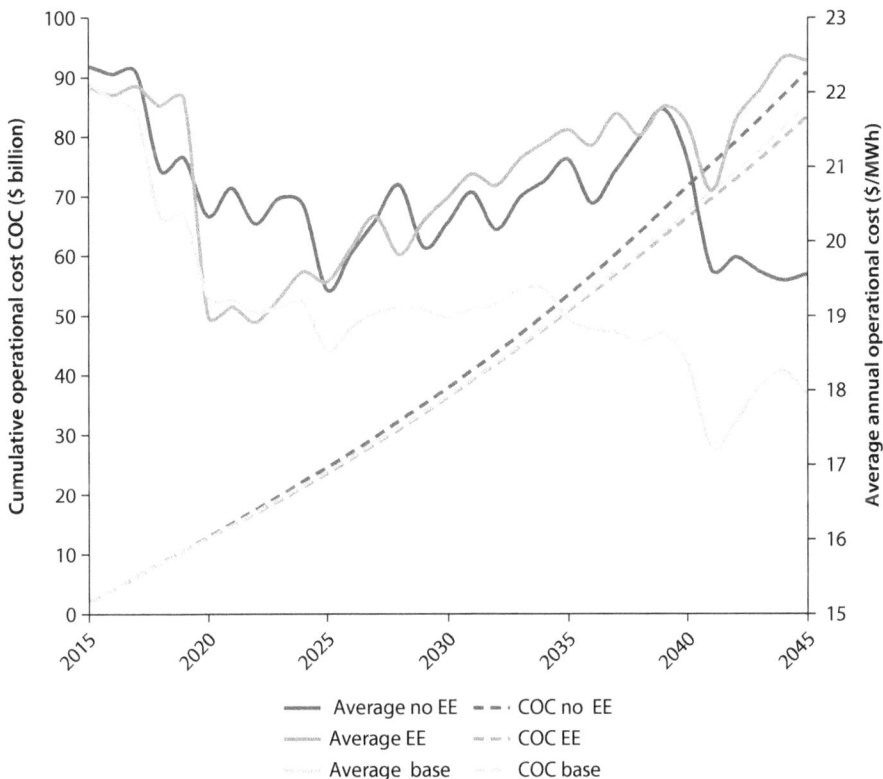

Note: COC = cumulative operational costs; EE = energy efficiency; MWh = megawatts/hour).

Stuck in Transition • http://dx.doi.org/10.1596/978-1-4648-0971-2

Figure 5.41 Cumulative Emissions for the Least-Cost and Base Case Scenarios, 2015–45

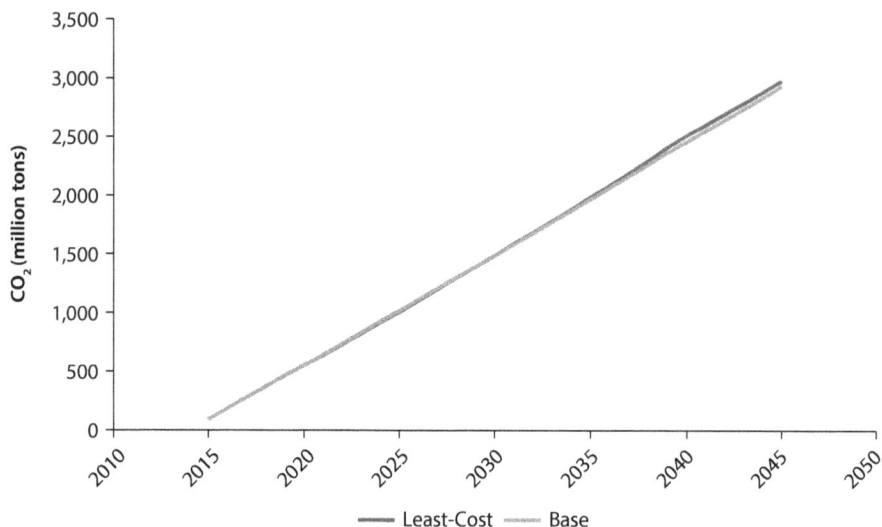

Note: CO_2 = carbon dioxide.

similar to those achieved from investing in about 20 GW of variable renewable energy technologies for a total overnight cost of about US$25 billion. The assumption for transmission costs is on the low side, although it could increase many times and still be less expensive than in the Base Case scenario.

Gasification of the Northern zone is uneconomical. It does not occur in the Least-Cost Case scenario, as expected.

(ix) *Sensitivity to natural gas price.* Given the uncertainty of the gas price, a separate sensitivity analysis of natural gas prices was performed, increasing the cost of natural gas in the Western zone from US$1.41/GJ to US$2.54/GJ. This is the same value as the cost of natural gas in the Southern zone. The higher cost of natural gas in the Western zone directly affects the generation and transmission plan; it results in a 29 percent increase in total coal-fired generation (from 2,100 to 2,700 TWh) and a 37 percent decrease in gas-fired generation (from 1,900 to 1,200 TWh). Increased gas prices make PV power economical earlier on, leading to an 85 percent increase in total PV production (from 115 to 211 TWh). Under higher gas prices, the West–South interconnection is no longer economical. A transmission capacity of 3 GW will be needed to interconnect the Northern and Western zones (compared with 7 GW in the original Least-Cost Case scenario).

Summary and Conclusions from the Least-Cost Analysis

Kazakhstan's power system is a rather low-cost system to operate (table 5.13). The LCOE across all scenarios ranges from US$31.1/MWh to US$41.5/MWh. In addition, Kazakhstan's current power planning is not far from the least-cost path.

Table 5.13 Summary of Scenario Costs

Cost	Base	Green [a,b]	Regional Export	Least-Cost[c]
Systemwide LCOE (US$/MWh)[d]	35.1	45.1 (33)[e]	34.8	31.1 (34.6)
Total undiscounted annualized CAPEX (US$ billions)	81.56	96.2	83.36	54.62 (50.8)
Total discounted annualized CAPEX (US$ billions)	25.3	28.9	25.5	17.4 (15.2)
Total operational cost (US$ billions)	85.3	82.4	93.6	91.1 (104.9)
Average operational cost (US$/MWh)	19.2	20.4	19.7	20.6 (23.2)
Total emissions (million tons CO_2)	2,932	2,460	3,252	2,977 (3,400)
Average emissions intensity (ton CO_2 per MWh-e)	0.69	0.64	0.71	0.69 (0.8)
CO_2 reductions from 2012 levels by end of optimization period (%)		40		

Notes: CAPEX = capital expenditures; CO_2 = carbon dioxide; LCOE = levelized cost of electricity; MWh = megawatts/hour; MWh-e = megawatt hours of electricity.

a. The cost of energy efficiency investments was not accounted for in any of the numbers presented in this table.

b. The cost of converting combined heat and power plant coal-fired units to gas-fired units was not accounted for in the calculations.

c. The values in parentheses represent the results for the variation of the Least-Cost scenario that considers the economic cost of natural gas.

d. A discount rate of 6 percent was assumed.

e. The value in parentheses is the LCOE if the benefit of global externalities associated with CO_2 savings were considered.

The Base Case scenario's systemwide LCOE is US$35.1/MWh, while the Least-Cost Case scenario's LCOE is US$31.1/MWh. Nevertheless, undiscounted annualized CAPEX requirements, ranging from US$54.6 billion (under the Least-Cost Case scenario) to US$96.2 billion (under the Green Case scenario), are needed to meet growing demand. Some investments have very important short-term implications, because the current investment focus on rehabilitation/retrofitting of current assets will help Kazakhstan's system increase its reserve margin above the current value of approximately 11 percent, improve efficiency in operations, and increase system security.

Insights from the results of the Base and Least-Cost Case scenarios suggest that Kazakhstan could follow a hybrid path that includes less investment in variable renewable energy technologies after 2030, while increased focus and funding are allocated for transmission assets to connect the three electricity zones. By that time, the country could be well-prepared for natural gas exploitation and allocation of large amounts of natural gas toward a gas-dominated rather than a coal-dominated power sector. Unifying the system and shifting toward natural gas will have significant environmental benefits due to CO_2 reductions and reduced coal-related pollutants (such as mono-nitrogen oxides, sulfur oxides, and particulate emissions).

The LCOE values do not include the costs of the energy efficiency program. If the Green and Base Case scenarios were compared based on undiscounted total costs, however, the Green Case scenario could only prove less expensive than the Base Case if global externalities were accounted for and the energy efficiency program cost US$75/MWh or less. If CHP conversion in the Northern zone does not take place, any energy efficiency program that costs US$105/MWh will render the Green Case scenario less expensive than the Base Case.

Increasing exports appears promising. At a total annualized undiscounted cost of US$1.6 billion, Kazakhstan can export up to 309 TWh over 2015 to 2045. The price of electricity in the region is, on average, US$50/MWh, which would result in revenues of US$15.45 billion. This amount more than recovers the additional investment requirements. The bulk of additional investments will be for coal-fired units in the Northern zone, with the remainder for gas-fired units—specifically CCGT units in the Western zone and OCGT units in the Southern zone. The investments in natural gas units in the Southern zone are needed, for the most part, to balance variable renewable energy production rather than for export purposes. Given the limited gas reserves and transmission infrastructure, however, the government's electricity export targets appear overly ambitious and can only be met with new gas-fired capacity, which may not be a viable option.

The Green Case scenario is the most expensive of all the scenarios, in LCOE (US$41.5/MWh) and capital requirements (US$96.2 billion). The Green Case scenario's LCOE is affected by the fact that variable renewable energy penetration is the highest for this scenario, thus raising capital costs for investing in flexible technologies with low capacity factors. In addition, the operational costs are high, compared with the other scenarios, because of the gasification of 1.6–1.8 GW of originally coal-fired CHP capacity in the Northern zone. However, the Green Case scenario becomes less expensive than the Base Case scenario if the global externalities associated with CO_2 emissions are accounted for and the energy efficiency program costs less than US$75/MWh.

The Green Case scenario has far less emissions than the other scenarios (400 million tons less than in the Base Case scenario) because of the implementation of an aggressive energy efficiency scenario that achieves a 24 percent reduction in peak demand by 2045. By the same year, CO_2 emissions will have been reduced by 40 percent. While the Base Case scenario meets national policy targets for variable renewable energy penetration, the Least-Cost Case scenario results in a nearly identical level of CO_2 emissions but costs US$27 billion less. As demonstrated by comparing the LCOE of each generator type provided in figure 5.42, high CAPEX renewables (wind, PV, and nuclear) are not forced on line and less expensive gas in the Western and Southern zones is exploited. Kazakhstan should strongly consider investing in domestic transmission to create a fully integrated national grid to use domestic resources more effectively. Considering the economic cost of natural gas, rather than the actual price, leads to a reduction of about US$4 billion in annualized CAPEX and an increase of about US$14 billion in OPEX.

Figure 5.42 Base Case Scenario 2020 Supply Curve: Generation (TWh) versus Levelized Cost of Electricity

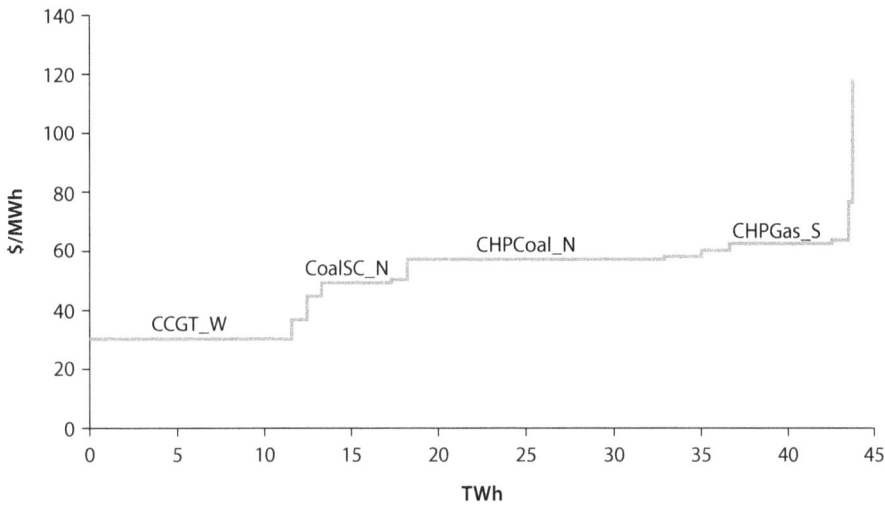

Note: CCGT = combined cycle gas turbine; CHP = combined heat and power plants; MWh = megawatt hours; N = North; S = South; SC = supercritical; TWh = terawatt hours; W = West.

Notes

1. The Asian Development Bank study limits its demand forecast for South Kazakhstan only because the study focuses on regional demand/supply and energy/water strategic issues with the Kyrgyz Republic, Tajikistan, and Uzbekistan.

2. In some cases, modified econometric models have been used for demand forecasting because of the lack of historical data. Fichtner GmbH & Co. used such an approach, where the demand drivers were estimated from the experiences of other countries and/or data for a few years for Kazakhstan (Fichtner GmbH & Co. 2012).

3. KEMA has used historical data and regression analysis to identify the GDP and value of industry as the most important indicators in its econometric load-forecasting model. It is interesting that population growth does not seem to correlate well with demand growth in Kazakhstan, according to historical data.

4. Alternative energy in KazEnergy's *National Energy Report 2013* comprises renewables, together with natural gas and nuclear power.

5. Decree of the President of the Republic of Kazakhstan in 2013.

6. At the time the analysis was performed, the exchange rate was US$1 = KZT 260.

7. The International Energy Agency lists uranium in the region at US$0.6/GJ. It was assumed here that it will be subsidized in Kazakhstan, similar to coal and gas, since it is a domestic fuel.

8. PV, wind, and hydro projects were mandatory (not subject to economic optimization) in the model, as per the renewable energy target specified in each scenario. However, their cost assumptions are shown.

9. That is, it is impossible for the software to determine the best capacity mix of solar and wind that will satisfy the energy target (for example, 10 percent by 2030). Thus, a defined capacity mix needs to be entered as input.

10. An approximately 610 kilometer gas pipeline from the South to the North will cost about US$1.35 billion; assuming a 50-year life, the transport cost of this gas is roughly US$0.56/GJ. With a gas price of US$2.54/GJ in the South, the price of gas in the North is assumed to be US$3.1/GJ.

11. The graph represents the capacity required to export electricity, assuming a 100 percent capacity factor. The actual capacity would be higher than the figures in the graph.

12. North–South transmission projects in the Least-Cost Case scenario differ from those planned (mandatory) to start operating in 2018, increasing the capacity by 750 MW.

13. The operational costs of renewables are assumed to be zero. The average operational system costs in US$/MWh are calculated by dividing total system operational costs (fuel + variable operations and maintenance) by total generation (thermal + renewables).

14. Figures were converted to 2015 U.S. dollars.

15. In general, the calculation of a systemwide LCOE requires systemwide inputs similar to the system OPEX and CAPEX figures, while a technology-related LCOE requires inputs similar to the OPEX, CAPEX, and capacity factors of a specific technology.

16. The cost estimates used cost data from the Alma transmission project in India. The overhead transmission lines are assumed to be at 500 kilovolts, which costs US$300,000/kilometer, with transfer capacity of 1,000 MW (the South–West line length is assumed to be 520 kilometers, while the North–West line is assumed to be 460 kilometers). The lengths of lines were approximated using current transmission network maps and Google Earth.

References

CEC (Center for Energy Economics). 2012. "Natural Gas Value Chain: Pipeline Transformation." Presentation.

DNV-GL. 2015a. "Integration of Renewables in Power Market and Power System, Kazakhstan."

DNV-GL. 2015b. "Regulatory and Institutional Improvement for Renewable Energy Investment in Kazakhstan."

Fichtner GmbH & Co. 2010. "CAREC Regional Master Plan."

———. 2012. "Central Asia Regional Economic Cooperation: Power Sector Regional Master Plan." Technical Assistance Consultant's Report. Asian Development Bank, Manila.

GoK (Government of Kazakhstan). 2014. "Energy Concept 2030." Concept on Development of the Fuel and Energy Complex of Kazakhstan until 2030.

IEA (International Energy Agency). (Ed.). 2013. Technology Roadmap: Wind Energy. Paris: OECD/IEA

———. (Ed.). 2014. Technology Roadmap: Solar Photovoltaic Energy. Paris: OECD/IEA

KazEnergy. 2013. "KazEnergy National Energy Report." Presentation (undated). Astana, Kazakhstan.

KEMA. 2010. "Roadmap for the Development of a Competitive Generation Market."

Spees, Kathleen, & Lester B. Lave. 2007. "Demand Response and Electricity Market Efficiency." *The Electricity Journal* 20(3): 69-85.

World Bank. 2014. "Social Value of Carbon in Project Appraisal. Guidance Note to World Bank Group Staff." World Bank, Washington, DC.

CHAPTER 6

Key Findings and the Challenges Ahead

Overview

Kazakhstan's power system has made considerable strides, and is reasonably well-functioning, especially compared with those of its Central Asian peers. Nonetheless, there remains a considerable reform agenda that is incomplete and requires action. There are various course corrections to be examined in an effort to move forward.

Progress for a more efficient power sector would have been more meaningful had the government of Kazakhstan not rolled back its achievements in liberalizing the wholesale market. The halting of the country's power generation privatization program has discouraged strategic foreign investment; the liquidity of the earlier thriving spot market has been seriously reduced; and excessive government control over the market has reemerged. Furthermore, sector regulation has not sufficiently advanced; rather, it has retreated in some areas (for example, the reintroduction of generation tariff regulation and the abolishment of the Agency for Regulation of Natural Monopolies).

An impending generation capacity deficit in the mid-2000s prompted the government to resort to excessive administrative controls and micromanagement of the sector, resulting in what is currently a semi-reformed state that is unattractive for investors. Foreign investment is crucial to overcome, in particular, capacity challenges. It is also required for modernizing distribution and improving the environmental performance of the power generation subsector.

Kazakhstan's power sector liberalization model continues to be sound, and there is no compelling reason for a profound paradigm shift. The decisive undertaking is for Kazakhstan to complete its reform agenda by not only addressing the unresolved issues of regional electricity cooperation within the Central Asian Power System but also taking various actions.

The government's latest strategic direction is well-aligned with the key findings of this study. Its successful implementation, however, will require focus and a sustained effort on the part of the government.

Key Findings

The findings of this study relate to capacity restriction and the security of Kazakhstan's power supply, with the aim toward environmentally sustainable development. The findings are primarily based on conclusions that derive from the results of the system least-cost modeling analysis, together with the recommendations of earlier studies and technical advice. Given the existing abundance of inexpensive domestic mineral resources and the lack of a well-defined and enforceable carbon pricing system, the analysis highlights the following:

- Kazakhstan's primary reliance on coal and gas remains economically justified, and the increased share of gas-fired generation (in regions with existing gas infrastructure) enhances system flexibility and contributes substantial environmental benefits, including emissions savings. The old coal-based generation infrastructure needs to be retired and replaced by supercritical technology in the Northern zone and gas-fired technologies in the Southern and Western zones.
- The government's strategy toward energy independence and its status as a substantial electricity exporter should be reviewed. Under the study's Regional Export scenario, the levelized cost of electricity (LCOE), capital expenditure, operating expenditure, and intensity of emissions are similar to those represented in the Base Case scenario. The price for exported electricity at generation should be U.S. 3.3 cents/kilowatt hour or more to achieve net gains. The Regional Export Case scenario requires additional investment in coal and gas—specifically in combined cycle gas turbine units in the Western zone and open cycle gas turbine units in the Southern zone—to meet system flexibility requirements. Given the limited gas infrastructure and proven available gas, the government's export targets appear to be overly ambitious and potentially should be scaled down.
- The government's strategy of gasification of the Northern zone is not economically justified. Full interconnection of Kazakhstan's power system into one unified system would substantially contribute to economic and environmental improvements. Therefore, the government should focus on gasification of the Southern zone and accelerated power system interconnection in the Northern, Southern, and Western zones.
- The government's efforts to ensure a much higher share of renewable energy in the generation mix is sound, although its alternative energy targets (30 percent by 2030 and 50 percent by 2045) may be overly ambitious. Unless global externalities are taken into account, the targets would result in a considerable economic cost. Therefore, the government should reassess its renewable energy targets.
- Reducing energy intensity contributes to substantial savings and environmental benefits to the end user; however, in the absence of quantified social and environmental benefits, the energy efficiency program would have to be less than US$13/megawatt hour (MWh) to be successful. The government should consider implementing an economywide energy efficiency program under this threshold and ensure a high return on energy efficiency measures.

- The Green Case scenario generates a reduction of 471 million tons of carbon dioxide (CO_2) emissions over the projected period. It shows that the intensity of emissions almost halves during the optimization period—from 0.98 tons of CO_2 per MWh in 2015 to 0.45 tons of CO_2 per MWh in 2045. This occurs as a result of Kazakhstan's system improving over time, as new, less costly, and more efficient technologies come on line. The decommissioning and rehabilitation of aged and inefficient coal- and gas-fired units should be continued.
- The government's ambitious renewable energy targets are inconsistent with its aim to contain future tariff increases to maintain its competitiveness for the electricity-intensive commodities that dominate its exports. Under the Green Case scenario, LCOE is US$41.5/MWh in comparison with US$35.1/MWh in the Base Case scenario and US$31.1/MWh in the Least-Cost Case scenario. Should the high renewable energy costs not be passed on to the end-user tariffs, it is difficult to envisage how the implicitly high subsidies to renewable energy generators will be funded.

Key Challenges

The key challenges identified under the government's latest strategy (especially Energy Concept 2030) largely correspond to the key results that emerge from this study—in particular, to ensure energy security by addressing underinvestment in generation and the systematic liberalization and development of competition. However, the indications to date suggest that the government intends to achieve these key objectives essentially by way of command-and-control methods and continued government micro-meddling. The risk is high of a departure from the government's declared energy strategy to promote competition and its least-cost expansion plan.

Kazakhstan's power sector faces various challenges. These are aggravated by plunging commodity prices and the consequent reduction of industrial production and power demand through

- High electricity intensity and generation capacity restriction without a cushion to prevent the risk of supply security;
- Formidable investment requirements; and
- Ineffective reversals of regulation and reform.

These challenges are interlinked. Some of the recommended solutions cut across them.

High Energy Intensity and Generation Capacity Constraints

Energy is used very inefficiently in Kazakhstan, reflecting the legacy of the former Soviet era. This contributes to the rapid growth in the demand for energy. Because Kazakhstan's economy is dominated by extractive industries and its products have low value addition, it is highly energy-intensive.

Kazakhstan ranks among the top 10 most energy-intensive economies in the world. It uses more than twice as much energy per unit of gross domestic product (GDP) compared with the average for Organisation for Economic Co-operation and Development (OECD) countries. Mirroring its high energy intensity, the country is among the most carbon-intensive economies in the world, and more than three times the average of OECD countries.

The generation capacity reserve margin stood at 53 percent in 2000, when peak demand reached its lowest level since the breakup of the former Soviet Union. The margin rapidly and steadily shrank to a dangerously low 4 percent in 2012. Since 2008, the combination of the global fiscal crisis and the generation tariff cap program has gradually reversed this downward trend. Although the reserve margin currently stands at 11 percent, to achieve an acceptable sense of system security for the long term will require consideration of new greenfield capacity.

Daunting Investment Requirements

The mobilization of investment is, indeed, an overwhelming challenge to overcome to implement the government's ambitious generation expansion plan under its Energy Concept 2030. The 7,500 MW of new capacity in 2013–30—estimated at approximately US$5.5 billion (US$325 million per year)—appears to be underestimated, relative to KazEnergy estimates and the World Bank's modeling results (figure 5.20). For 2015–30, KazEnergy shows an investment requirement of approximately US$54 billion (US$3.6 billion per year), while the World Bank's Base Case scenario approximation is US$42 billion (US$2.8 billion per year).

Although it may have attracted some much-needed funding, the government's seven-year generation investment stimulation program (2009–15) represents only a small fraction of the overall investment amount required. This raises concerns about the efficiency, transparency, and long-term sustainability of the state program.

To meet growing demand, undiscounted annualized capital expenditure requirements range from US$54.6 billion (Least-Cost Case scenario) to US$96.2 billion (Green Case scenario) over 2015–45. This represents a significant level of investment, equivalent to 0.8 percent (Least-Cost Case scenario) and 1.4 percent (Green Case scenario) of 2013 GDP (US$231.9 billion).

Inefficient Regulation and Reversal of Reforms

Despite the healthy progress achieved in the late 1990s and early 2000s, Kazakhstan's power sector remains highly inefficient, operationally and environmentally. Although at times resulting in short-term gains, the government's decision since the mid-2000s to reverse earlier sector reforms has substantially aggravated longer-term prospects. These have worsened the climate for investment, damaged competition, and squeezed out the private sector.

The conceptual basis of Kazakhstan's power sector liberalization model on the one hand—including that outlined in the Energy Concept 2030—continues to be sound. On the other hand, although there is no compelling reason for a paradigm shift, it is clear that the reform agenda needs to be completed by

- Halting the ongoing shift to oligopolization and excessive state control;
- Reversing renationalization by reprivatizing much of Samruk-Energy's generation assets;
- Improving the sector's poor investment climate with transparency and independent regulation;
- Placing the sector on a sustainable long-term path by completing the multimarket model (that is, bilateral, spot, balancing, ancillary services, and capacity market); and
- Focusing the state's role on strategy and policy making in lieu of micromanaging the competitive market segments of the sector, monitoring and countering excessive market power, and introducing incentive- and performance-based regulation.

(i) *Generation.* The creeping renationalization process that has recently accelerated under the vigorous acquisition strategy of state-owned Samruk-Energy (the energy subsidiary of the Samruk-Kazyna national sovereign fund, which owns nearly half of the country's generation capacity) potentially risks the preservation of competitive conditions in the sector. During 2008–11, Samruk-Energy acquired considerable generation assets, including the two largest generation plants. In 2013, the installed generation capacity of the company was 9,667 MW (47 percent of the national total), generating 33.5 terawatt hours of power (37 percent of the national total).

The fact that Samruk-Energy and the system operator (Kazakhstan Electricity Grid Operating Company, KEGOC) are owned by the same parent company (Samruk-Kazyna's multisector national holding) further raises the risk of excessive market power at the wholesale level. The company's extremely high net profit margins—at 20 percent in 2013 and 43 percent in 2012—may be indicative of its unmitigated market power. In a truly competitive market, net profits would settle at a much lower level.

Three major power plants of national importance (Ekibastuz GRES-1, Ekibastuz GRES-2, and Aksu) account for approximately half of the power volume traded on the national wholesale market. With regional power markets within the country de facto closed to meaningful competition, the contestable segment of the bulk wholesale market has rapidly evolved toward an oligopolistic structure.

(ii) *Transmission.* KEGOC, as the system operator, has carried out a large-scale rehabilitation and extension investment program. As a result, the performance of the national high-voltage network has improved considerably.

Although government legislation ensures equal access to the transmission grid for all qualified wholesale market participants, in practice the access regime remains immature and lacking in clear and detailed procedures, protocols, and transparency. It is also badly in need of a credible dispute resolution mechanism.

Over the past two decades, portions of the Kazakhstani transmission grid—in particular, the North-South interconnector—have become increasingly congested, requiring a new generator to avert the risk of output not necessarily finding its optimum market. This situation contributes to the overall unattractive investment climate of the power sector. The fact that a portion of Kazakhstan's transmission grid is owned by vertically integrated utilities adds another source of uncertainty in addition to allegations of biased access preferences for own generation.

KEGOC provides ancillary services under nontransparent arrangements, bundled into a generic charge imposed on all grid users. This limits the ability of generators to engage in providing such services. The transmission grid is thus open to some transactions—primarily standard sell/buy deals—but closed to others, such as ancillary services.

(iii) *Distribution.* Kazakhstan's distribution system is excessively fragmented, with nearly 200 distribution entities. As a result, distribution is the weakest link in the power sector value chain, in addition to often nontransparent ownership, politically influenced tariff setting, grossly insufficient funds for modernization because of the compressed distribution margins, and inefficient operation. These problems are reflected in the high share of outdated equipment and excessively large network losses.

Final consumers have contractual relations only with electricity supplier organizations (ESOs) that can neither own nor operate the low-voltage equipment that belongs to the regional electricity distribution companies (RECs). The latter do not interface directly with consumers and have no contractual obligation to address complaints. As a consequence of the division in responsibility between RECs and ESOs, the final end users often remain unprotected and inadequately served.

Anecdotal evidence suggests that there are significant problems with the quality and reliability of power supply, especially in rural regions. No systematic database exists to record the quality and reliability of the supply at the retail level. Furthermore, distribution companies are not required to apply internationally accepted measures to assess supply reliability.

(iv) *Market structure.* The government's sector strategy aims, as its objectives, to ensure energy independence and ease capacity restriction in generation. Indications to date, however, suggest that these objectives are to be achieved essentially by way of command-and-control tactics. The risk remains high of a departure from the government's energy strategy to promote competition and its least-cost generation expansion path.

- *Bilateral contract market.* Following independence, the wholesale market was fully liberalized based on bilateral sale/purchase transactions. In a major setback in 2008, however, the state imposed major restrictions, capping generation prices and banning generator-to-generator and supplier-to-supplier trades.

- *Balancing market.* A real-time balancing market was fully designed, but its launch has been repeatedly delayed since 2008 for reasons that do not appear credible, given the recent marked slowdown in demand growth and the new additions to flexible load-following generation capacity.
- *Ancillary services market.* There is no organized market for ancillary services.
- *Spot market.* This was launched in 2004, and by 2007 had reached an 18 percent market share. However, the spot market's growth has been undermined by government decisions, including a ban on trader participation and inter-ESO transactions in the spot market.

The existing multimarket model is incomplete and not fully functional; the earlier well-functioning submarkets (that is, bilateral contracts and spot) were weakened by excessive state controls and interventions since the mid-2000s; and the ancillary services market is hand-controlled by KEGOC in a nontransparent manner. The introduction of a much-needed balancing market has been long delayed.

The recent drive toward horizontal and vertical integration in generation raises legitimate concerns about the potential abuse of market power. In addition, although a generation capacity market has been designed, it is considerably flawed and raises concerns that its effectiveness may be compromised following its significantly delayed launch in 2019.

Recommendations

The policy recommendations here aim to improve Kazakhstan's regulatory environment, ensure the completion of its market reforms, and attract critical private investment. The government of Kazakhstan must address the lack of investor interest in the sector by creating an enabling environment that is stable, is transparent, and falls within a predictable legal and regulatory framework. Given the long lead times to build new generation facilities, the government should reduce regulatory uncertainty with several rapid and specific measures, including the following:

(i) *Generation*

- Contain or reverse the ongoing horizontal state-ownership concentration and oligopolization of generation assets to preserve reasonably competitive conditions in the sector. Mitigating market power is particularly essential amid a continued tight supply situation.
- Contain or reverse the ongoing renationalization of generation assets within Samruk-Energy by divesting all—or at least the controlling stakes—of its subsidiary generators that are deemed not to be strategic to professional investors. This would be in line with a new, ambitious national privatization program for 2016 and beyond.

- Introduce a credible and well-designed generation Capacity Market to attract, transparently and competitively, qualified investors to address the underlying capacity deficit.
- In conjunction with the launch of the Capacity Market, phase out the existing generation price cap scheme (recently extended to 2022), because it is profoundly inconsistent with the core rationale of the Capacity Market. Coexistence of the administered tariffs and the Capacity Market may compromise the latter's integrity.
- In the combined heat and power plant sector, eliminate the cross-subsidization of heat production by power generation, which makes this sector uncompetitive in the wholesale electricity market, as well as unattractive to investors.

(ii) *Transmission.* Develop detailed, fully transparent, and legally binding open-access network rules, incorporating international best practice standards, as part of the electricity Grid Code by

- Providing full transparency about the real-time availability of transmission capacity;
- Ensuring equal access amid transmission constraints by way of a transparent, bid-based mechanism to enable all users to compete for scarce transmission capacity, and establishing a simple and efficient dispute resolution mechanism; and
- Enforcing, legislatively, the neutrality of the system operator (KEGOC) in the face of all sellers and buyers, and ensuring that KEGOC operates the grid strictly according to competition-neutral, but security-conscious, protocols. However, this requirement may be counter to KEGOC's current subsidiary status within the Sovereign Wealth Fund, Samruk-Kazyna national holding, whose other subsidiary, Samruk-Energy, is the largest owner of power generation assets and the principal player in the wholesale market.

(iii) *Distribution.* Although the government has acted assertively in the generation sector in the face of a looming generation crunch, its attention to the distribution sector—the critical weak link in the electricity value chain—has been inadequate. The government has leveraged—albeit relatively inefficiently—considerable capacity modernization and expansion in generation, but the distribution sector remains mired in a poor technical and financial setup, largely because of inefficient tariff regulation. This does not allow for an adequate investment component in the highly compressed distribution margin. Therefore, this study recommends the following:

- Effectively and legally unbundle the trading, supply, and other business (generation and distribution) of the RECs. Put in place strict and monitorable cost and functional separation rules (ring fencing) to prevent distributors from favoring their legally separate supply affiliates—otherwise, retail competition will continue to exist largely only on paper.

- Encourage independent suppliers and traders to enter the REC- and ESO-dominated retail market to create meaningful competition for consumers.
- Consolidate the extremely large number of inefficient distributors into a smaller set of commercially viable entities.
- Design and develop smart grids; mandate the installation of automated commercial metering/communication devices.

(iv) *Wholesale market reforms*

- Preserve the bilateral contracts market. Mostly because of the market inconsistency with least-cost-based generation, a progressively higher share of wholesale trade should be channeled into more transparent, bid-based trading mechanisms, such as the spot market (Kazakhstan Operator of Electric Power and Electric Energy, KOREM) and the balancing market.
- Put in place a well-functioning and liquid trading floor, such as a spot market—which is essential for integrating variable renewable energy into the wholesale market. The government should restore the normal functioning of the spot market by lifting unnecessary administrative and legal restrictions.
- Launch the live operation of the much-delayed balancing market. The risk of possible hikes in the balance price should be managed by the sector regulator, capping it at a reasonable level.
- Transform the existing KEGOC-administered mechanisms of selected ancillary service purchases into an organized, stand-alone, bid-based procurement system under appropriate regulatory oversight. The cost of these services, incurred by the system operator, should be clearly shown in the unbundled transmission tariff structure.
- Introduce a well-designed Capacity Market to stimulate generation expansion on a competitive and transparent basis. The conceptual design currently under consideration should be reassessed to prevent the unnecessarily high complexity of running two submarkets (short- and long-term); inconsistent use of administrative price caps along with a bid-based Capacity Market; and excessive financial risk concentration at the balancing market operator (KEGOC) under its Capacity Purchase Agreements with investors.
- Abolish the generation tariff cap scheme upon the operational launch of the Capacity Market.

(v) *Sector regulation*

- Review the overall institutional framework for regulation, with the primary objective of developing a streamlined and efficient regulatory system to attract critical new investment and international expertise to the power sector.

- Strengthen regulatory capacity and increase regulatory autonomy to attain credibility among market participants. The regulator's governance structure (for example, the terms of employment of key staff, structure of the executive board, and budget) should be updated in line with best practice. Recent hosting of the sector regulator within the Ministry of National Economy should be reassessed as inconsistent with enhanced regulatory autonomy.
- Introduce across the electricity value chain incentive- and performance-based regulations to replace the inefficient cost-of-service regulation and tariff cap plus investment commitment scheme.
- Phase out cross-subsidies between electricity and heat production in combined heat and power plants, because the latter are largely uncompetitive in the wholesale power market.
- Establish vigilant market monitoring and effective controls on market power across the entire sector value chain, given the much-increased vulnerability of the wholesale system to market power.
- Carry out a detailed tariff structure assessment that may reveal the need to rebalance by aligning regulated retail tariffs with the cost of wholesale power. The regulator should avoid retail rate freezes that expose distributors to an unsustainable squeeze on their cash flow when rising wholesale costs approach (or possibly exceed) fixed retail rates.

Reference

GoK (Government of Kazakhstan) 2014. "Energy Concept 2030." Concept on Development of the Fuel and Energy Complex of Kazakhstan until 2030.

Generation Tariff Caps by Groups of Power-Generating Companies, 2016–18

(tenge/kilowatt hour, excluding value-added tax)

Group	Tariff caps on electricity		
	2016	2017	2018
Group 1 LLP "Ekibastuz GRES-1", named after B.Nurzhanov JSC "Ekibastuz Power Station GRES-2" JSC "Eurasion Energy Corporation" (Aksu GRES)	8.8	8.8	8.8
Group 2 JSC "Zhambyl GRES," named after T.I. Batyrov	8.7	8.7	8.7
Group 3 JSC "Astana-Energia" (CHP 1 and 2) LLP "Karaganda Zhyly" (Karaganda CHP-1,3) JSC "Pavlodarenergo" (Pavlodar CHP-2,3, Ekibastuz CHP) JSC "AES Ust-Kamenogorsk" CHP JSC "Arselor Mittal Temirtau" (CHP-2, CHP-PVC)	7.5	7.5	7.5
Group 4 LLP "Kazakhmys Energy" (Karaganda GRES-2, Balkash CHP, Zhezkazgan CHP) JSC "Alyminiy Kazakhstan" (Pavlodar CHP-1)	6.0	6.0	6.0
Group 5 LLP "SevKazEnergo" JSC "Ridder" CHP JSC "SSGPO" CHP	8.05	8.05	8.05
Group 6 LLP "AES Sogrinskaya" CHP LLP "Bassel group" LLP (Karaganda GRES-1) LLP "Stepnogorsk" CHP	8.3	8.3	8.3
Group 7 JSC "Atyrau" CHP JSC "Aktobe" CHP JSC "Tarazenergocentr"	7.3	7.3	7.3

table continues next page

Group	Tariff caps on electricity		
	2016	2017	2018
Group 8 State-owned public enterprise, "Kentau Servis" LLP "Kazzink-TEK" (Tekeliskaya CHP, Karatalskaya HPP) LLP "Shachskteploenergo"	7.5	7.5	7.5
Group 9 State-owned public enterprise, "Arkalyk" CHP State-owned public enterprise, "Kostanai Teplo-Energia Company" JSC "Zhaykteploenergo"	7.6	7.6	7.6
Group 10 JSC "Almatinskue electricheskue stanzii" (CHP-1, CHP-2, CHP-3, Kapchagai HPP, Kaskad HPP)	8.6	8.6	8.6
Group 11 LLP "MAEK-Kazatomprom" (CHP-1, CHP-2, CHP)	15.32	15.32	15.32
Group 12 LLP "Zhanazhol GTPP" JSC "Aktobe Ferro Alloy Plant" Kazchrom (Akturbo) LLP "Uralsk GTPP" LLP "Kristall Management"	8.8	8.8	8.8
Group 13 Bukhtarma HPP LLP "Kazzink" LLP "AES Ust-Kamenogorsk HPP" LLP "AES Shulbinsk HPP" JSC "Shardara HPP"	4.5	4.5	4.5
Group 14 LLP "Aktobe Rail and Beam Plant"	10.64	10.64	10.64
Group 15 JSC "3-Energoortalyk" (Shymkent CHP-3) State-owned public enterprise, "Kyzylorda teploelectrozentral"	8.3	8.3	8.3
Group 16 Moinak HPP	8.8	8.8	8.8

Note: Tariffs reported in the table were approved by Order No. 160 of the Ministry of Energy of Kazakhstan as of February 27, 2015.

CHP = combined heat and power plant; GTPP = gas thermal power plant; HPP = hydropower plant; JSC = Joint Stock Company; LLP = Limited Liability Company.

Least-Cost Expansion Model Description

Model Inputs and Outputs

The modeling used OptGen and Stochastic Dual Dynamic Programming (SDDP), which are two computational tools developed by Power System Research (PSR). OptGen is used for least-cost expansion plans (generation transmission and interconnections) of multiregional power systems. It can be integrated with SDDP, which is a detailed transmission-constrained scheduling model. To run SDDP, the technical specifications of generators and transmission lines, demand data, and variable costs are fed into the model. The main inputs for OptGen include the capital expenditure costs of a group of possible future projects. SDDP returns the dispatch scheduling that fits the least-cost expansion capacity plan, which is optimized in OptGen (figure B.1).

Basic Mathematical Formulation of Generation Capacity Expansion Problems

The goal (objective function) of the least-cost expansion model is to minimize total system costs. Total system costs are the sum of fixed and variable costs during the optimization period (table B.1). Typical fixed costs include the capital expenditure cost of new generation as well as fixed operations and maintenance costs. Variable costs include the cost of fuel, variable operation and maintenance costs, and the cost of carbon, if applicable.

In its simplest form, the objective function of the generation capacity expansion models is

$$Minimize \sum_{y=y1}^{NY} \sum_{g=g1}^{NG} \sum_{t=t1}^{NT} \sum_{f=f1}^{NF} (Cost_{FIXED}\ g,y + Cost_{VAR}\ g,f,y,t)$$

(B.1)

where t, g, y, are indexes:

$$ti\ Time\ increments\ [Hrs]\ 1 < i < T\ and\ \sum_{i=1}^{T} (ti) = 8760$$

Figure B.1 Inputs and Outputs of the PSR Expansion Planning Software

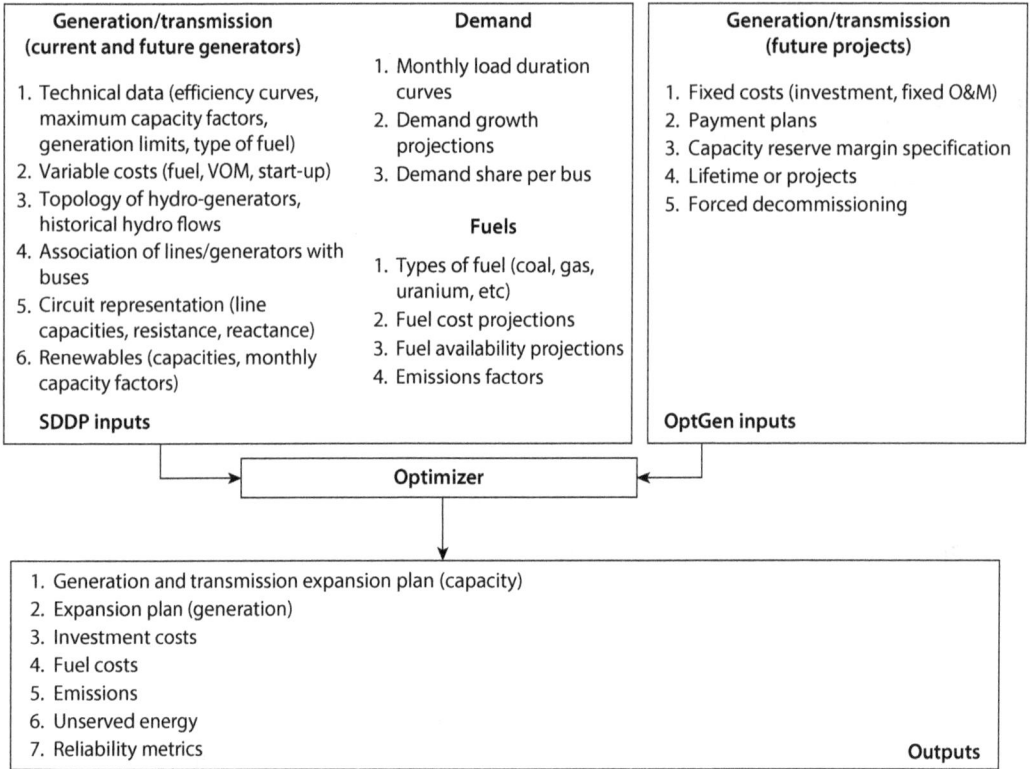

Generation/transmission (current and future generators)	Demand	Generation/transmission (future projects)
1. Technical data (efficiency curves, maximum capacity factors, generation limits, type of fuel) 2. Variable costs (fuel, VOM, start-up) 3. Topology of hydro-generators, historical hydro flows 4. Association of lines/generators with buses 5. Circuit representation (line capacities, resistance, reactance) 6. Renewables (capacities, monthly capacity factors) **SDDP inputs**	1. Monthly load duration curves 2. Demand growth projections 3. Demand share per bus **Fuels** 1. Types of fuel (coal, gas, uranium, etc) 2. Fuel cost projections 3. Fuel availability projections 4. Emissions factors	1. Fixed costs (investment, fixed O&M) 2. Payment plans 3. Capacity reserve margin specification 4. Lifetime or projects 5. Forced decommissioning **OptGen inputs**

Optimizer

1. Generation and transmission expansion plan (capacity)
2. Expansion plan (generation)
3. Investment costs
4. Fuel costs
5. Emissions
6. Unserved energy
7. Reliability metrics   **Outputs**

Note: O&M = operations and maintenance; SDDP = Stochastic Dual Dynamic Programming.

Table B.1 Breakdown of Fixed and Variable Costs for the Least-Cost Expansion Plan

Fixed	*Variable*
Capital expenditures for new generators	Fuel cost
Fixed operating and maintenance	Variable operating and maintenance
	Cost of carbon (if applicable)

$$gn: Generator\ 1 < n < NG$$

$$ym: Year\ 1 < m < NY$$

$$fl: Fuel\ 1 < l < NF$$

and

NT: *total number of time increments that constitute 1 year*

NG: *total number of generators*

NY: *total number of years that constitute the optimization period*

NF: *total number of fuels*

$Cost_{FIXED}$ g, y: *fixed cost per generator per year (Annualized CAPEX and fixed O&M) [$]*

$Cost_{VAR}$ g, f, y, t: *variable cost per generator per fuel consumed at increment t and year y [$]*

Typically, optimization problems require the implementation of a set of constraints that put limits on the acceptable region in which a solution can exist. Such constraints are mathematical equations that describe laws of physics (similar to the supply and demand balance) or actual system constraints (for example, the regional availability of fuel), and they are either applied automatically or activated by the user in the case of PSR software. A typical set of constraints for generation capacity models includes

- Supply and demand balance;
- Reserve margin constraint (if applicable);
- Generation within technical limits of generators;
- Fuel reserve constraints;
- Association of projects (for example, one project can enter only to replace another);
- Thermal energy supply and demand balance (for combined heat and power plants) and
- Power flow constraints (when transmission is included).

Kazakhstan's National Electricity Grid

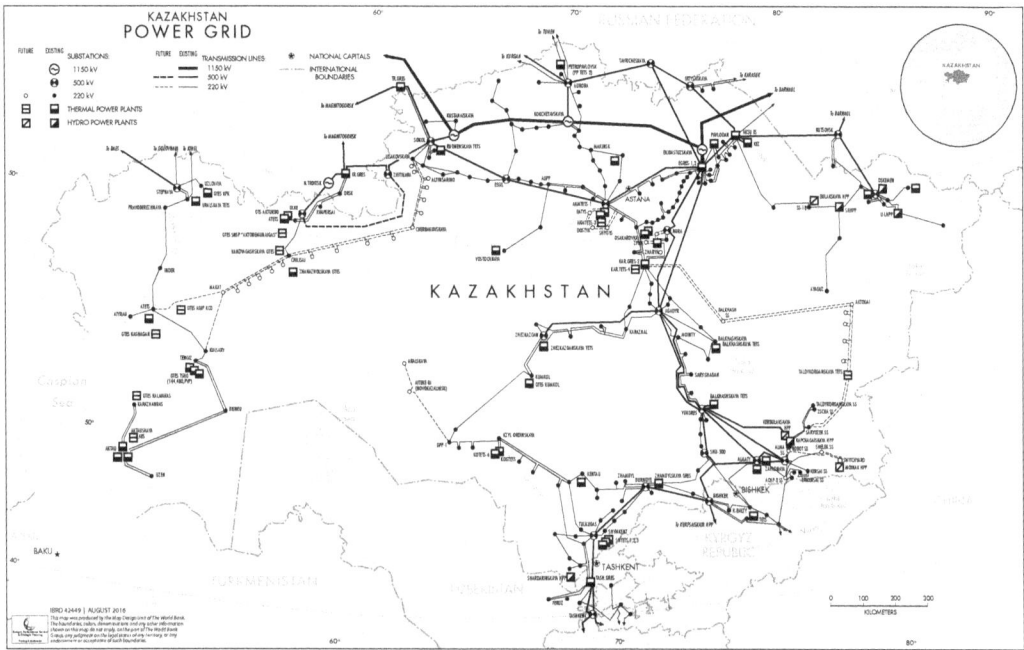

Source: Kazakhstan Electricity Grid Operating Company.

Environmental Benefits Statement

The World Bank Group is committed to reducing its environmental footprint. In support of this commitment, we leverage electronic publishing options and print-on-demand technology, which is located in regional hubs worldwide. Together, these initiatives enable print runs to be lowered and shipping distances decreased, resulting in reduced paper consumption, chemical use, greenhouse gas emissions, and waste.

We follow the recommended standards for paper use set by the Green Press Initiative. The majority of our books are printed on Forest Stewardship Council (FSC)–certified paper, with nearly all containing 50–100 percent recycled content. The recycled fiber in our book paper is either unbleached or bleached using totally chlorine-free (TCF), processed chlorine–free (PCF), or enhanced elemental chlorine–free (EECF) processes.

More information about the Bank's environmental philosophy can be found at http://www.worldbank.org/corporateresponsibility.